AF886280

SOTHIS PRESS

Sothis Press aims to completely reimagine the public domain book with an emphasis on universal access to captivatingly designed print editions. We believe that technology will enable every single book ever published to be perpetually accessible in beautiful print form. Sothis is beauty in ubiquity.

www.sothis.press
1 Waterhouse Square, London, ECIN 2ST

The Daily March of Transpiration in a Desert Perennial

Edith Bellamy Shreve

SOTHIS PRESS

ISBN 978-1-0368-0889-1

This edition is published by Sothis Press © 2023. The company is registered in England and Wales. The Company Number is 14900389.

THE DAILY MARCH OF TRANSPIRATION IN A DESERT PERENNIAL

BY

EDITH BELLAMY SHREVE

WASHINGTON, D. C.

PUBLISHED BY THE CARNEGIE INSTITUTION OF WASHINGTON

1914

CARNEGIE INSTITUTION OF WASHINGTON
Publication No. 194

PRESS OF GIBSON BROTHERS, INC.
WASHINGTON, D. C.

CONTENTS.

THE DAILY MARCH OF TRANSPIRATION IN A DESERT PERENNIAL.

INTRODUCTION.

A consideration of the various types of perennial plants indigenous to the vicinity of the Desert Laboratory at Tucson, Arizona, brings to view some striking differences between the species concerned, in respect to their total annual water-losses. The perennials fall into the general types, succulent and non-succulent, the latter being easily classified further into three physiological groups, namely: (1) small plants whose perennial parts are confined to roots or underground stems; (2) plants continuously in leaf; (3) tropophytic plants. Examples of the second class are *Covillea tridentata, Lycium berlandieri, Celtis crassifolia, Encelia farinosa, Hyptis emoryi;* examples of the third class are *Prosopis velutina, Acacia greggii, Fouquieria splendens, Jatropha cardiophylla,* and *Parkinsonia microphylla.* The tropophytic plants are alike in that they all drop their leaves more or less quickly in times of drought, but *Parkinsonia microphylla* differs from the others in having a continuous covering of epidermis over all of its parts, even the limbs over 100 years old having an active chlorphyll layer covered with an unbroken living epidermis. Thus, although the trees are without leaves during times of drought, they must still lose a large quantity of water. Some of the evergreen plants have a continuous epidermis, but these are all much smaller than *Parkinsonia* and hence expose a smaller evaporating surface. In spite of this necessity for a large loss of water, Spalding* calls *Parkinsonia* a "highly successful desert species, growing in abundance and equally well on slopes of all exposures." Shreve† found a high death-rate in seedlings of *Parkinsonia,* but reached the conclusion that the critical period in the life of the plant is over after the first two years.

The abundance of *Parkinsonia* on hill slopes in an arid region is an index of its apparent success in spite of the exposure of a large evaporating surface during the entire year and the difficulty its seedlings have during the first two years. In consequence of the ability of *Parkinsonia* to overcome these adverse conditions, it was selected as the subject of transpiration-behavior studies. A general study of transpiration behavior of seedlings and adult plants during different seasons of the year is in progress, but certain fluctuations in the daily transpiration rate make a separate publication on this phase of the work advisable. Only such description of the plant, its habitat, and its anatomical features will be given as seems to apply either directly or indirectly to the phase of transpiration mentioned.

*Spalding, V. M., Distribution and movements of desert plants, Carn. Inst. Wash. Pub. 113, 1909.

†Shreve, F., Establishment behavior of palo verde, Plant World, vol. XIV, p. 294, 1911.

The work is being carried on at the Desert Laboratory at Tucson, Arizona, and is made possible through the kindness of Dr. D. T. MacDougal, who has placed the facilities of the laboratory at the writer's disposal.

Parkinsonia microphylla is a leguminous tree with minute deciduous leaves. Plate I*C* shows the general appearance of the adult tree, while No. 4 of plate I*B* gives an idea of a seedling which has grown under natural conditions for a year. No. 5 of the same figure is a seedling, about eight years old, which was killed in transplanting. All adult trees in natural conditions carry from one to eight dead limbs and have from 5 to 10 per cent of the medium-sized branches dead; about 30 per cent of the twigs are dead for a distance of 2 to 5 cm. from the ends, and after the first dry season succeeding growth practically all twigs are dead for a distance of 0.2 to 0.5 cm. from the ends. (See also Shreve, F., *loc. cit.*) The above is merely an estimate from general observation and is not based on statistics.

The leaves appear in the late winter after the rains, are usually shed during the arid fore-summer, reappear within a few days after the summer rains begin, and persist for varying lengths of time, according to the amount of rain falling during the early autumn. During the autumn of 1910 the leaves had all fallen from the trees on Tumamoc Hill before the end of September, but in 1912 many leaves persisted until killed by frost in January of 1913. The rain records taken at the Desert Laboratory show a much drier autumn for 1910 than for 1912. During both these years a tree near the laboratory, where it probably received water from an artificial source, retained its leaves until January.

Observation of the trees on Tumamoc Hill showed the following general response to progressive desiccation of air and soil: First, the leaflets drop, leaving the rachises still attached; then if no rain comes the rachises are shed, and after a month or so the ends of the twigs die. All leaves do not necessarily fall at once, for sometimes as many as half of the leaflets may remain for a month after the others have fallen. These observations were made throughout 1910 and the summer, autumn, and winter of 1912. At the same time, observations were made on two plants which probably received an abnormal supply of water. The first was a tree situated near a south door of the laboratory, where its roots probably received a reserve supply of water from beneath the laboratory porch. This tree has had no new dead twigs or limbs for the last three years at least, and during those years has retained its leaves from four to six weeks after the other trees were bare. The second was a plant about 3 feet high, whose main trunk had been cut off in the spring of 1909. Its root was thus unusually large in proportion to its top, and it was otherwise under unnatural conditions, since it was shaded by the laboratory during a large part of the day and the soil about its roots was covered with a mulch of crushed stone. This plant has lost no twigs or branches since the cutting took place and has leaves throughout the year, except in mid-winter.

During the above period many trees were observed both on Tumamoc Hill and on the foot-hills about Tucson for evidence of the place where dying

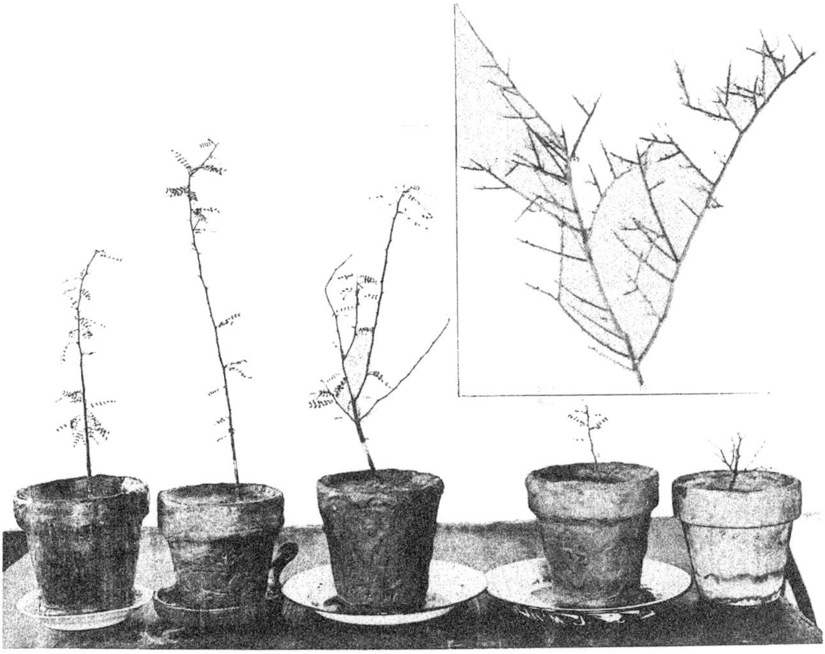

A. Parkinsonia microphylla, end branch of a tree.

B. Plants of Parkinsonia used in experiments VI-X. Counting from left to right, Nos. 1-3 were raised in the green-house, Nos 4-5 were grown under natural conditions having been transplanted to pots a short time before the experiments. Nos. 1-4 are one year old. No 5 is about eight years old.

C. Parkinsonia microphylla, adult tree.

begins. Chlorophyll is abundant in all parts of the bark, except at the short base of the main trunk, so that the death of any part of the tree is easily detected by a change in color of the bark from bright green to brown. Certain dead twigs were marked, and in the following year the dead region was found to have extended farther down the stem, but no case was observed where dying had begun below the tip. After a limb is dead it is of course impossible to tell the stages by which it died, but the number of dead twigs and short dead branches existing on live branches makes it seem probable that death begins at the tips and progresses down the stem gradually. The death of an entire limb seems in no way to affect the health of the plant as a whole.

As is the case with most of the Leguminosæ, the plant possesses leaf-movements. During darkness the leaflets are "closed"—that is, they stand with their dorsal surfaces in parallel planes about 0.1 cm. apart—and with the coming of daylight the leaflets separate, making various angles with the rachis, the "wide-open" position being when the leaflets lie in the same plane on opposite sides of the rachis. During cool or moist seasons the leaflets remain "open" or partly open during the day, but during the dry seasons of spring and fall they close within half an hour after sunrise and remain in this position all day, sometimes opening for about half an hour near sunset. Leaflets from potted plants that have been well watered open wider as a rule in the morning than do those from adult trees and are slower in closing. If, however, the potted plants are not given water for several days the movements are more like those of the leaflets of adult trees.

The size of leaflets varies, both in area and thickness, with the age of the plant and with the soil and atmospheric conditions existing during the period of their growth. On seedlings from one to eight years old the area of one side of a leaflet averages 0.04 sq. cm.; on an adult tree the average is 0.02 sq. cm.; and on hot-house-grown seedlings a year old it is 0.06 sq. cm. The leaflets which come from very small branches emanating from the main trunk near the base are frequently about the same size as those on young plants. The diameters are approximately in inverse ratio to the area. Figure 1 shows the diameters and arrangement of tissue in leaves of different types. The different drawings are on the same scale and were made with a camera lucida.

The following differences of structure may be concerned with the transpiration studies which follow. The section of the leaflet from an adult tree (*A*, fig. 1) shows large epidermal cells with heavily thickened walls, stomata about midway down the first row of cells, very few intercellular spaces, palisade cells and stomata on both sides of the leaf. The section from a plant grown all its life in the green-house (*C*, fig. 1) shows no thickening of the epidermal cells, many large intercellular spaces, palisade cells on the dorsal side only, and stomata on both sides. The section from the leaf grown in the green-house for a year and then placed out of doors for six weeks (*B*, fig. 1) shows a structure in several ways intermediate between the other two. The epidermal cells are thickened nearly as much as in the section from the adult tree; palisade cells occur on one side only, stomata on both sides, and the intercellular spaces are fewer and smaller than on the plant kept in the green-house all its life, but are larger and more frequent than on the one from the tree. The section from the twig of the tree (*D*, fig. 1) shows epidermal cells with heavily thickened walls, from three to four subepidermal cells with walls very heavily thickened, stomata sunken into the first and second layers of subepidermal cells, and guard cells of the stomata covered with a heavy layer of cuticle. The section from a

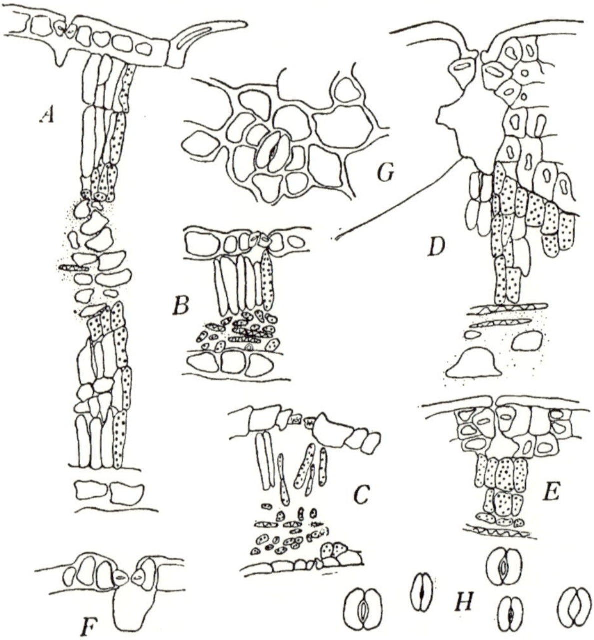

FIG. 1.—Camera lucida drawings of leaf and stem sections of *Parkinsonia microphylla*.

A. Cross-section of leaflet from adult tree.
B. Cross-section of leaflet from plant grown in green-house for a year and then placed in the open for 6 weeks.
C. Cross-section of leaflet from plant kept in green-house all of its life.
D. Tangential section of epidermis of twig from adult tree.
E. Tangential section of twig from hot-house-grown plant.
F. Cross-section of stoma from rachis of adult tree.
G. Surface view ot stoma from leaflet of adult tree.
H. Stomata from leaflets showing various apertures.

twig of a green-house-grown plant (*E*, fig. 1) shows epidermal cells thickened, but there are no subepidermal cells and the intercellular spaces are larger and more frequent than in the section from the tree. An examination of many sections showed that the stomata of the rachis are sunken slightly (*F*, fig. 1)—that is, that they appear always near the base of the epidermal cells, that the stomata of the leaflet itself sometimes appear near the bottom of this layer, but that the usual place is near the top of the layer, and that the stomata of the branches are always sunken below the layer of epidermal cells.

In general it may be said that the evaporating surfaces, both external and internal, are largest in the plant which has carried on its entire development in the green-house, next in the plant which has been in the green-house for one year after germination, then in the open for six weeks, and least in the adult tree. Of the different parts of the tree examined, the leaflets lose water by evaporation most easily, the rachises next, and the branches least easily.

TRANSPIRATION STUDIES.

METHODS.

The best method yet devised for measuring transpiration is doubtless the determination, by weighing, of losses sustained by sealed potted plants. In the case of *Parkinsonia* this method was not practical, since the plants raised in the green-house differ from those growing in the open, both in leaf-structure and in the size of stems and leaves. The influence of the two environments on the size of seedlings is shown in plate I *B* and on the size and structure of the leaves in figure 1. It proved almost impossible to transfer small seedlings from their natural position in the ground to pots, there being only one survival out of some thirty which were transplanted. It thus seemed necessary to use a method for the measurement of water-loss *in situ*. Cannon's method* is adequate for a rough estimate of comparative water loss from different plants under similar conditions, and seems to have several advantages over Stahl's† method, yet for the purposes of the present study, which require accurate quantitative measurements of water loss under natural conditions, a more accurate method had to be sought. The older methods, based on the absorption of moisture by means of sulphuric acid or calcium chloride, have many of the same objections as the polymeter method, *i. e.*, the plant is not under normal conditions of humidity, wind, and sunshine; moreover, the gain in weight of the absorbing material may not represent the actual amount of water given off by the plant, but is equal to this amount plus or minus the loss or gain of moisture in the air of the bell-jar during the experiment. If the bell-jar is a large one and the part of the plant under the jar is losing a small amount of water, this error may amount to as much as 100 per cent.

A method was finally worked out which contains some of the features of both Cannon's method and the absorption methods. By a reference to fig. 2 it will be seen that the plant or part of the plant to be measured passes through a platform and is covered by a bell-jar, within which are an open dish of calcium chloride, a maximum thermometer, and a dew-point apparatus, the last being so arranged that the dew-point within the jar may be taken without any disturbance of the conditions within the jar.

---

*Cannon, W. A., A new method of studying the transpiration of plants in place, Bull. Torr. Club, vol. XXXII, p. 515.

†Stahl, E., Einige Versuche über Transpiration und Assimilation, Bot. Zeit., III, p. 117, 1894.

The details of the platform are shown in *B* of fig. 2. The groove seen there was filled with mercury and thus an air-tight joint was made possible. In the earlier experiments this groove was not used, but the bell-jar was sealed to the platform with plastocene. The mercury seal proved to be a great improvement with respect to speed of operation and the certainty of securing an air-tight joint. The surface of the platform was made non-absorbent by some 15 coats of shellac. The part of the plant to be used was led into the platform through a small hole in a rubber stopper which fitted tightly into the circular hole. The stopper was flush with the bottom of the groove, so that mercury could pass over the joint.

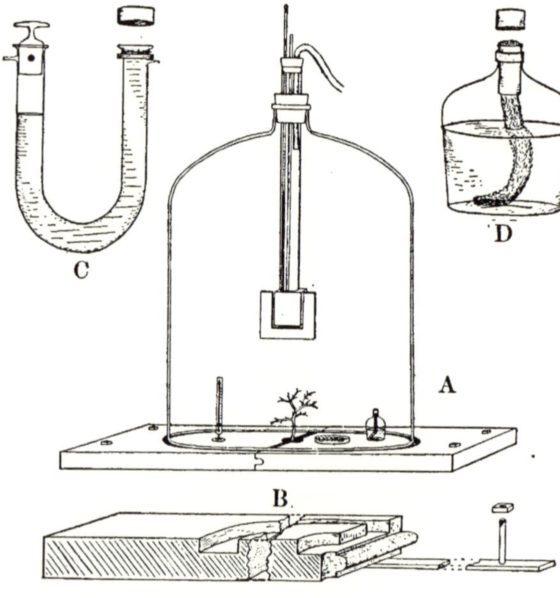

FIG. 2.—*A*. Apparatus used for measuring transpiration *in situ*.
B. Details of the base used in *A*.
C and *D*. Forms of atmometers used in *A*.

The dew-point apparatus was made for the purpose and is a modification of Allouard's form. It consists of a metal cube 2 cm. on a side, one face of which is accurately planed, nickeled, and highly polished. Close to, but insulated from this face, is a polished metal flange for comparison. Into the metal cube is fastened a glass tube of about 1.5 cm. diameter, which passes through a rubber stopper in the top of the jar, the tube being long enough so that the metal cube may be placed at any desired height within the jar. From the top of the jar project a 0.1 degree certified thermometer and the tubes necessary for the entrance and exit of ether. The ether was conducted through a rubber tube about 15 feet long before it was allowed to escape into the air. In the experiments carried on in 1911 ether was used to cool the metal surface, but in 1912 carbon dioxide* was led into the tube from a high-pressure tank. By this means the dew-point was obtained much more quickly than with ether and the danger of physiological disturbances caused by waste ether was avoided. Frequent tests with this dew-point apparatus showed that the temperature of appearance and disappearance of moisture differed less than 0.1° C. when the temperatures were above −4° and not more than 0.2 degree when the temperatures were below −4°.

The calcium chloride was exposed in shallow weighing-bottles, the covers of which were air-tight when the salt was not in use. Merck's *C. P.* calcium chloride was used, but to guard against the absorption of an appreciable amount of carbon dioxide from the air by traces of potassium hydroxide which might be present, carbon dioxide gas was made from calcium carbonate and sulphuric acid, and passed dry through the calcium chloride until saturation resulted.

Finally the entire jar was shaded, except a small opening which allowed a beam of light to fall on the whole plant. Constant illumination was maintained by changing the position of the opening every 10 minutes, so that no shadow ever fell on the plant. By this method of shading it was found easy to keep the temperature of the air within the bell-jar within one or two degrees of that on the outside, provided the volume of the jar was at least 1,000 times that of the plant and the jar was opened and fresh air let in every two hours.

The amount of water given off by the plant was found from the gain in weight of the calcium chloride plus or minus a correction ascertained in the following way. The dew-point was taken at the beginning and at the end of the period and from the Smithsonian tables the weight of water in a cubic meter of saturated air at the temperature of the two dew-points was found. The difference between these two weights was then multiplied by the volume of the jar, proper corrections being made for the space occupied by the objects under the jar. Corrections for altitude were made in a few cases where the error from neglecting the correction amounted to more than 0.001 gm.

The following conditions are provided in the present method: (1) The water lost by the plant can be measured accurately to milligrams by means of well-constructed weighing-bottles and an accurate dew-point apparatus; (2) the duration of the experiment may be from 15 minutes to 2 hours; (3) transpiration readings in sunlight may be obtained by means of a special method of shading; (4) the air temperature and humidity within the jar may be made to closely approximate the condition of the outside air by a proper manipulation of the shading device and of the amount of calcium chloride exposed; (5) the influence of air currents can not be measured.

In order to measure the evaporation rate of a less complex surface under similar conditions and thus to secure data regarding the behavior of the plant itself, a second bell-jar was set up, identical with the first one, except

* The use of carbon dioxide as a cooling agent was suggested by Dr. B. E. Livingston.

that the plant was replaced by a small atmometer which lost about the same amount of water as the plant. This atmometer (*D*, fig. 2) consisted of a tube containing a wick which dipped below into a small reservoir of water and ended above in sand, slightly piled. A cover was provided for the evaporating surface for use when the atmometer was removed to be weighed. This atmometer was standardized to an open pan of water in a dark room and its loss reduced to loss per square centimeter of water surface. The results of this standardization brought out the fact that the atmometer's rate was a fairly constant one under the same conditions; for the average of three readings of 24 hours each it varied less than I per cent from the extreme readings. The actual losses from the atmometer during the experiments were always reduced to unit area and the loss per unit area from the plant for the same period was divided by this number, the result being called the relative transpiration.* It might seem on *a priori* grounds that the shape of the curve obtained from these results would be parallel to a true curve, but that the actual values might be either too high or too low, since the conditions under the two bell-jars were seldom absolutely identical; yet the values for relative transpiration obtained in 1912, when atmometer and plant were under the same bell-jar, are in such close agreement with these earlier ones that the results obtained by the use of two bell-jars are shown to be as accurate as are those obtained when one jar was used.

While this last method has the advantage over the earlier one that plant and atmometer are under identical conditions, it has the disadvantage that a larger beam of light must be let into the jar, with the consequent greater rise in temperature within the jar. It will be seen, from the tables of September 1912, that the temperatures under the bell-jar rose every day as much as 3° above the outside air temperatures. As in the case of humidities, it is true here that the plant is not placed under unusual conditions, unless by chance the hottest day of the year was selected for the experiment, and this was not the case. Yet, when only parts of a plant are used, it seems best that the conditions surrounding the branch under experimentation should be as nearly as possible identical with those around the rest of the tree. In the subsequent tables, temperatures and humidities are given in order that any difference may appear which might affect conclusions. A number of experiments were discarded where the meteorological conditions within and without the bell-jar differed over 10 per cent; but no experiments were discarded which might bring out negative evidence for the main conclusion.

Great care was taken to have all joints tight during the experiment and to keep the weighing-bottles and atmometer air-tight when not under the bell-jar. Before each set of experiments trials were made in which the atmometer was used as a plant and its losses determined by weight and compared with the gain in weight of the calcium chloride plus or minus the correction determined from the dew-points. Sensitive chemical balances were used, but readings were taken only to 1 mg. and the error taken as plus or minus

*Livingston, B. E., Carn. Inst. Wash. Pub. 50.

0.5 mg. In the final results no important conclusions are drawn from numbers obtained from differences in weighings smaller than 10 mg.

The curves are all drawn to a scale which shows only readings which are surely significant and well out of the range of experimental error. The solid horizontal lines show the actual time and duration of the intervals measured and the broken lines connect the mid-points of these. This scheme is used because the irregularity of the time required for taking the dew-points made it impossible to take readings at stated intervals. The time was always read to the nearest minute and the hourly rate calculated. It will be noted that the experiments made in 1912 show a greater agreement in the time of readings, an advantage gained by the use of the carbon dioxide for cooling the dew-point apparatus.

The extremely small size of the leaves prevented the use of any of the usual methods for determining their area, and so a direct method was used as follows. From 10 to 20 leaflets of approximately the same size were taken at a time, placed carefully with their long diameters in a straight line, and their aggregate length read to 0.1 mm. These leaflets were then turned so that their short diameters could be measured in the same way. All of the leaflets used in an experiment were thus measured and the average long and short diameters were found. The leaflets were then treated as ellipses, which they are so far as the eye can tell. The area of an average leaflet was calculated and was then multiplied by the total number of leaflets. The rachises were placed end to end and their total length was read. Their average diameter was found from many readings with a vernier caliper. The rachises were treated as cylinders, the groove which exists along the top being ignored. The stems were treated in the same way as the rachises, readings of diameters in this case being taken at regular intervals of 1 cm., as closely as the presence of thorns and branches would permit. Live thorns were treated as branches and all dead ends of branches were left out of account. Small errors doubtless occur in the operation of this method, but the consistency of the results goes to show that the errors may be disregarded and that the areas determined are comparable.

For the sake of comparison, plants raised in the green-house from seeds and then placed in the open for six weeks were sealed in their pots and their water loss was determined by weighings. The evaporation rates in these cases were determined by the Livingston type of atmometer.* Atmometers which had been standardized under Dr. Livingston's direction were used and the proper corrections for reduction were made by using the coefficients furnished by him. The rates were reduced to unit area of water-surface in order that the relative rates might be compared with those obtained from trees by the bell-jar method. This standardization was done in a dark constant-temperature room, an open petri dish, with a reservoir to keep the water level constant, being employed. Two of the cups were run with an open dish in this manner and the other cups were reduced to the open-water equivalents by the use of the coefficients referred to above.

*Livingston, B. E., Carn. Inst. Wash. Pub. 50.

Unless the plants have some special physiological response to wind action the relative transpiration rates of potted plants ought to be comparable to the relative rates of the plants taken *in situ* in still air. A general observational record of wind velocity was kept and it was found that while a marked increase in wind velocity was always accompanied by an increase in the actual transpiration rate, this increase did not appear in the relative transpiration, thus showing that the plant was affected in the same manner as the atmometer. The few conclusions drawn in this paper from a comparison of the relative rates of plants *in situ* with potted plants are of course open to the objection that there may exist an unknown physiological response on the part of the plant to wind which did not appear.

EXPERIMENTATION.

All of the plants or branches used in the following experiments were in normal condition and were situated in exposed positions, so that the sun could fall upon them the entire day. Whenever over-night readings were taken the jar was completely covered with a black cloth. It will be seen that, as the experiments progress in chronological order, the agreement of meteorological conditions of the jars with each other and with the outside become better, due to the growing experience of the operator. A small amount of calcium chloride was exposed for the night readings, the amount being about tripled for the first reading in sunshine and then increased gradually as the day advanced and decreased in proportion as the afternoon progressed. In all cases it was necessary to spend one or two days in trials before the amount could be properly adjusted for a particular plant. Also, it frequently became necessary to take into account the changes of dew-point in the outside air. With the temperature kept within a degree or two of that of the outside air by means of the method of shading mentioned above, and with the dew-point held near that of the outside air, the humidity varied only a small percentage from the atmospheric humidity. In no case were the differences between bell-jar and air conditions greater than the differences existing between the atmospheric conditions of any two days. The best results were obtained when periods of one to two hours were used; in the long over-night periods results were never satisfactory. Of course two-hour periods might have been used at night as easily as in the day, if the purposes of the experiment had required it. It was thought advisable to eliminate so variable a factor as the wind and so no attempt was made to get readings in the wind except in the case of potted plants.

In the tables which follow, unless otherwise defined, T represents the transpiration rate per hour per square centimeter of area; E the evaporation per hour per square centimeter; and T/E the relative rate. All losses are given in fractions of a gram. Unless stated otherwise all readings were taken in bright sunshine. In table 1, the original readings and some of the details of calculation are given, while for economy of space these details are omitted in the succeeding tables.

Experiment I.

The subject of this experiment was a small 1910 leafless seedling, growing on the north slope of Tumamoc Hill. The area was found to be 7.99 sq. cm.

TABLE 1.—*Transpiration of a leafless seedling (A). Experiment I.*

Date.	Time.	Dew-point.	Gain in weight of calcium chloride.	Correction.	Total loss from plant.	Loss per hour from plant.	Loss per unit area *T*.	Time.	Loss per hour from atmometer.	Loss per unit area from atmometer. *E*	$\frac{T}{E}$
1911.		° *C.*									
Mar. 6.. Mar. 7..	5h00m p.m. 6 00 a.m.	6.4 −10.5	0.111	−0.057	0.054	0.0041	0.00052	4h40m p.m. 6 10 a.m.	0.0088	0.007	0.077
	7 00 a.m. 8 07 a.m.	5.8 3.0	.048	− .012	.036	.032	.00404	6 45 a.m. 8 00 a.m.	.024	.019	.213
	8 35 a.m. 10 35 a.m.	7.0 7.0	.091	.000	.091	.045	.00563	8 30 a.m. 10 25 a.m.	.044	.035	.162
	10 55 a.m. 1 00 p.m.	7.0 5.9	.058	− .005	.053	.025	.00319	11 00 a.m. 1 05 p.m.	.065	.052	.061
	1 50 p.m. 3 50 p.m.	2.9 6.0	.033	+ .013	.046	.023	.00288	1 45 p.m. 3 55 p.m.	.050	.040	.072

Date.	Time.	Dew-point.			Temperature.			Humidity.		
		Plant.	Atmometer.	Air.	Plant.	Atmometer.	Air.	Plant.	Atmometer.	Air.
1911.		° *C.*	° *C.*	° *C.*	° *F.*	° *F.*	° *F.*	*p. ct.*	*p. ct.*	*p. ct.*
Mar. 6..	5h 00m p.m.	6.4	6.4	6.4	62	62	63	51	51	51
Mar. 7..	6 00 a.m.	−10.5	−8.0	7.2	43	43	43	27	34	80
	7 00 a.m.	5.8	5.8	5.8	54	54	54	64	64	64
	8 07 a.m.	3.0	3.3	6.7	72	72	71	30	30	38
	8 35 a.m.	7.0	7.0	7.0	76	76	76	35	35	35
	10 35 a.m.	7.0	7.1	7.0	81	81	80	28	28	28
	10 55 a.m.	7.0	7.0	7.0	80	80	80	28	28	28
	1 00 p.m.	5.9	6.0	4.0	84	84	83	24	24	22
	1 50 p.m.	2.9	2.9	2.9	84	84	84	18	18	18
	3 50 p.m.	6.0	5.8	6.0	82	81	80	25	25	25

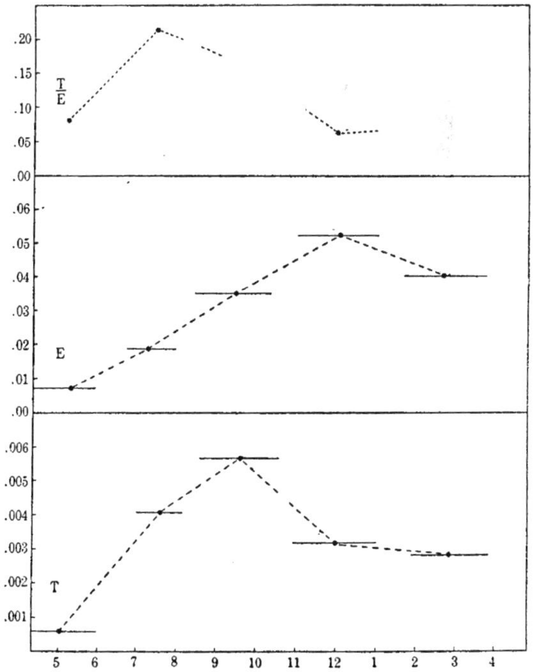

FIG. 3.—Graphs for transpiration of *A*, a small leafless seedling. Exp. I.

Experiment II.

Subject, a small leafless plant growing near the subject of experiment I. Area, 35.65 sq. cm. Age, approximately six to ten years.

Table 2.—*Transpiration of a leafless seedling (B). Experiment II.*

Date.	Time.	T	Date.	Time.	E	T/E
1911. Mar. 7... Mar. 8...	$5^h 18^m$ p.m. 6 07 a.m.	0 00022	Mar. 7.. Mar. 8..	$5^h 20^m$ p.m. 6 10 a.m.	0 00636	0.034
	6 52 a.m. 7 53 a.m.	.00073		6 55 a.m. 8 10 a.m.	.0111	.065
	8 05 a.m. 11 40 a.m.	.00454		8 35 a.m. 11 45 a.m.	.0334	.136
	12 00 m. 2 10 p.m.	.00539		12 25 p.m. 2 15 p.m.	.0525	.102
	2 50 p.m. 4 50 p.m.	.00252		2 45 p.m. 4 45 p.m.	.0382	.066
Mar. 9...	5 20 p.m. 6 20 a.m.	.00025	Mar. 9..	5 15 p.m. 6 30 a.m.	.00636	.039
	6 55 a.m. 7 55 a.m.	.00143		6 45 a.m. 8 00 a.m.	.0135	.106
	8 05 a.m. 10 05 a.m.	.00412		8 30 a.m. 10 09 a.m.	.0342	.120
	10 25 a.m. 12 25 p.m.	.00474		10 20 a.m. 12 20 p.m.	.0469	.101
	12 50 p.m. 2 50 p.m.	.00485		12 45 p.m. 2 45 p.m.	.0517	.094
	3 22 p.m. 5 20 p.m.	.00199		3 10 p.m. 5 10 p.m.	.0310	.064

Date.	Time.	Dew-point.			Temperature.			Relative humidity.		
		Plant.	Atmometer.	Air.	Plant.	Atmometer.	Air.	Plant.	Atmometer.	Air.
1911.		° C.	° C.	° C.	° F.	° F.	° F.	p. ct.	p. ct.	p. ct.
Mar. 7...	$5^h 18^m$ p.m.	2.5	2.5	2.5	67	67	67	..	..	..
Mar. 8...	6 07 a.m.	−6.5	−1.0	3.0	48	48	48	32	49	67
	6 52 a.m.	3.5	3.5	3.5	52	52	52	60	60	60
	7 53 a.m.	3.0	5.5	4.3	60	[1]66	60	44	41	48
	8 05 a.m.	4.5	[2]5.5	4.5	62	[2]70	62	44	37	44
	11 40 a.m.	16.0	14.0	7.0	84	84	82	46	40	28
	12 00 m.	6.5	6.5	6.5	82	[3]86	82	25	23	25
	2 10 p.m.	−7.5	−6.5	5.8	90	90	87	7	8	21
	2 50 p.m.	5.1	5.1	5.1	84	84	84	22	22	22
	4 50 p.m.	−0.3	−0.5	5.1	78	78	76	19	17	29
	5 20 p.m.	4.0	4.0	4.0	78	78	78	..	..	..
Mar. 9...	6 20 a.m.	−3.5	−1.0	2.5	56	56	56	..	..	..
	6 55 a.m.	2.2	2.2	2.2	56	56	56	47	47	47
	7 55 a.m.	7.8	3.5	3.5	71	71*	70	40	30*	29
	8 05 a.m.	3.5	3.5	3.5	72	72	72	29	29	29
	10 05 a.m.	12.0	11.7	6.0	85	84	84	36	35	24
	10 25 a.m.	6.2	6.2	6.2	86	86	86	20	20	20
	12 25 p.m.	14.5	13.8	6.0	95	96	93	29	27	18
	12 50 p.m.	1.8	1.8	1.8	94	94	94	12	12	12
	2 50 p.m.	−2.2	−2.0	1.8	96	94	93	9	9	13
	3 22 p.m.	1.8	1.8	1.8	92	92	92	14	14	14
	5 20 p.m.	3.0	3.0	3.2	78	79	76	24	22	25

[1] $8^h 10^m$ a. m. [2] $8^h 35^m$ a. m. [3] $12^h 25^m$ p. m.

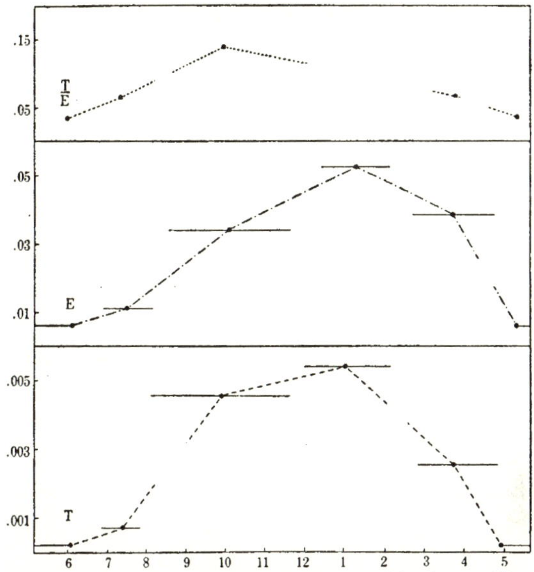

FIG. 4.—Graphs for transpiration of *B*, a small leafless seedling, March 8. Exp. II.

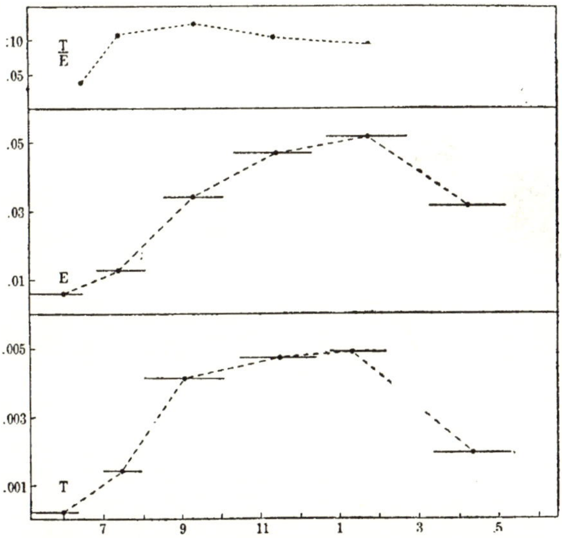

FIG. 5.—Graphs for transpiration of *B*, March 9. Exp. II.

EXPERIMENTS III TO V.

The subject of these experiments was an adult tree designated *C*, about 20 feet high, growing southwest of the laboratory in an exposed place. Three separate small branches were used, branch *c* being without leaves, while *d* and *e* had a normal number of healthy leaves. The areas were: *c*, 35.47 sq. cm.; *d*, 18.13 sq. cm.; *e*, 8.20 sq. cm. The platform was used as before, but was mounted on a framework arranged to support the jar, etc. The leaves on the tree were always completely closed before the readings were begun.

TABLE 3.—*Transpiration of a leafless branch (c) of an adult tree. Experiment III.*

Date.	Time.	T	Date.	Time.	E	$\frac{T}{E}$
1911. Mar. 11... Mar. 12...	5^{h} 20^{m} p.m. 6 45 a.m.	0.00031	Mar. 11... Mar. 12...	5^{h} 15^{m} p.m. 6 55 a.m.	0.006	0.049
	7 20 a.m. 7 55 a.m.	.00216		7 10 a.m. 8 05 a.m.	.015	.144
	8 33 a.m. 9 30 a.m.	.00575		8 25 a.m. 9 20 a.m.	.038	.151
	9 55 a.m. 11 00 a.m.	.00598		9 50 a.m. 11 10 a.m.	.045	.132
	11 25 a.m. 12 35 p.m.	.00338		11 35 a.m. 12 45 p.m.	.049	.068
	1 10 p.m. 2 10 p.m.	.00380		1 00 p.m. 2 00 p.m.	.050	.075
	2 30 p.m. 3 30 p.m.	.00292		2 20 p.m. 3 20 p.m.	.047	.062
	3 45 p.m. 4 45 p.m.	.00259		3 35 p.m. 4 40 p.m.	.029	.088
Mar. 13...	5 15 p.m. 7 25 a.m.	.00019	Mar. 13...	5 10 p.m. 7 30 a.m.	.0071	.026
	8 05 a.m. 9 25 a.m.	.00354		7 55 a.m. 9 15 a.m.	.022	.158
	9 30 a.m. 10 55 a.m.	.00434		9 35 a.m. 10 50 a.m.	.038	.113
	11 15 a.m. 12 35 p.m.	.00310		11 10 a.m. 12 40 p.m.	.044	.069
	1 00 p.m. 2 00 p.m.	.00382		1 05 p.m. 1 55 p.m.	.049	.078
	2 20 p.m. 3 20 p.m.	.00279		2 25 p.m. 3 15 p.m.	.044	.063
	3 45 p.m. 4 45 p.m.	.00330		3 50 p.m. 4 50 p.m.	.053	.062

TABLE 3—Continued.

Date.	Time.	Dew-point.			Temperature.			Relative humidity.		
		Plant.	Atmometer.	Air.	Plant.	Atmometer.	Air.	Plant.	Atmometer.	Air.
1911.		° *C.*	° *C.*	° *C.*	° *F.*	° *F.*	° *F.*	*p. ct.*	*p. ct.*	*p. ct.*
Mar. 11..	5h 20m p.m...	−3.5	−3.5	−3.5	59	59	59	28	28	28
Mar. 12..	6 45 a.m...	−7.5	−7.4	−1.5	40	42	40	40	37	65
	7 20 a.m...	−2.0	−2.0	−2.0	42	42	42	57	57	57
	7 55 a.m...	3.0	2.5	−1.5	58	56	56	47	49	37
	8 33 a.m...	−1.0	−1.0	−1.0	78	77	78	18	17	18
	9 30 a.m...	−1.5	−2.0	−1.5	80	81	80	15	14	15
	9 55 a.m...	−1.5	−1.5	−1.5	80	80	80	15	15	15
	11 00 a.m...	−3.0	−2.5	−3.0	85	85	83	12	12	14
	11 25 a.m ..	−3.0	−3.0	−3.0	84	84	84	12	12	12
	12 35 p.m...	−5.5	−4.8	−6.0	84	85	84	10	11	9
	1 10 p.m...	−7.0	−7.0	−7.0	85	85	85	9	9	9
	2 10 p.m...	−4.5	−5.5	−7.0	89	88	87	9	9	9
	2 30 p.m...	−7.5	−7.5	−7.5	86	86	86	9	9	9
	3 30 p.m...	−7.5	−6.5	−4.5	86	86	84	9	9	9
	3 45 p.m...	−4.5	−4.5	−4.5	82	82	82	10	10	10
	4 45 p.m...	−8.0	−7.5	−3.0	80	80	78	9	9	15
	5 15 p.m...	−3.0	−3.0	−3.0	72	72	72	18	18	18
Mar. 13..	7 25 a.m...	−1.3	−3.0	2.8	60	60	58	32	28	45
	8 05 a.m...	2.0	2.0	2.0	60	59	60	41	42	41
	9 25 a.m...	7.0	6.0	1.0	70	68	70	40	40	27
	9 30 a.m...	0.8	0.8	0.8	70	70	70	27	27	27
	10 55 a.m...	−5.0	−4.5	0.6	79	80	79	13	12	20
	11 15 a.m...	0.5	0.5	0.5	81	81	81	18	18	18
	12 35 p.m...	−5.0	−4.8	−1.8	85	85	83	10	9	14
	1 00 p.m...	−2.0	−2.0	−2.0	84	84	84	13	13	13
	2 00 p.m...	−5.0	−5.2	−3.0	90	90	89	9	9	10
	2 20 p.m...	−3.0	−3.0	−3.0	90	90	90	10	10	10
	3 20 p.m...	−7.0	−5.5	−2.5	85	87	85	9	9	12
	3 45 p.m...	−0.5	−0.5	−0.5	79	79	79	18	18	18
	4 45 p.m...	−3.5	−5.0	1.0	82	80	76	13	12	22

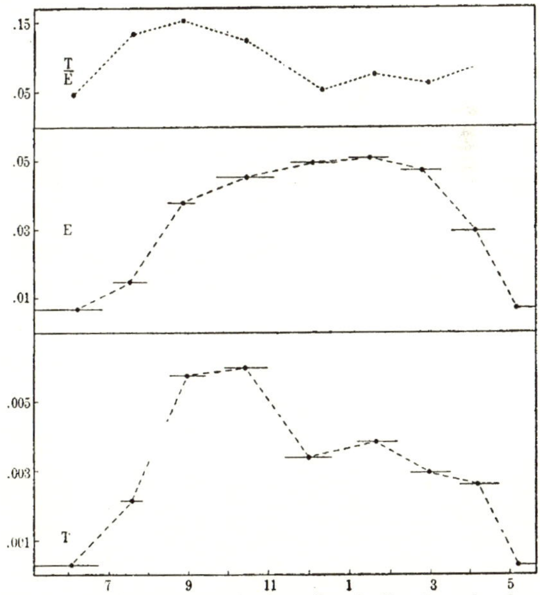

FIG. 6.—Graphs for transpiration of *c*, leafless branch of an adult tree, March 12. Exp. III.

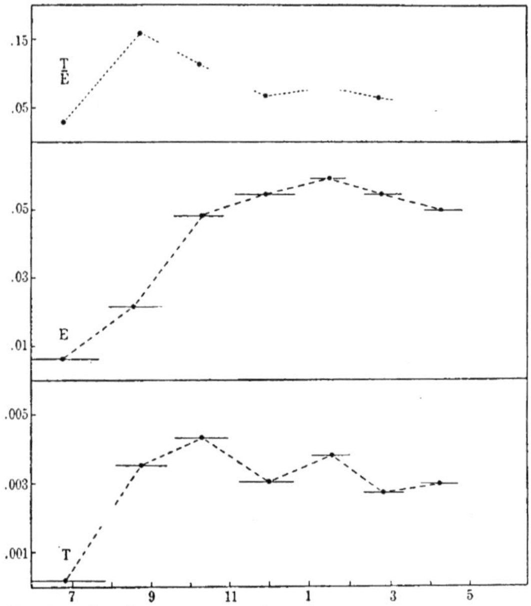

FIG. 7.—Graphs for transpiration of *c*, March 13. Exp. III.

TABLE 4.—*Transpiration of a branch in leaf (d) of an adult tree. Experiment IV.*

Date.	Time.	*T*		Time.	*E*	*T/E*
Mar. 13. Mar. 14.	5h 17m p.m. 6 19 a.m.	0.00007	Mar. 13. Mar. 14	5h 15m p.m. 6 17 a.m.	0.010	0.007
	6 55 a.m. 8 00 a.m.	.00571		6 58 a.m. 7 50 a.m	.018	.319
	8 25 a.m. 9 29 a.m.	.00988		8 22 a.m. 9 20 a.m.	.028	.353
	9 36 a.m. 10 55 a.m.	.00970		9 32 a.m 10 48 a.m.	.040	.241
	11 00 a.m. 12 25 p.m.	.00510		11 04 a.m. 12 18 p.m.	.049	.104
	12 33 p.m. 1 55 p.m.	.00904		12 35 p.m. 1 48 p.m	.052	.174
	2 04 p.m. 3 05 p.m	.00446		2 05 p.m. 2 55 p.m.	.049	.091
	3 22 p.m. 4 23 p.m.	.00331		3 25 p.m. 4 15 p.m.	.022	.....

TABLE 4—Continued.

Date.	Time.	Dew-point.			Temperature.			Relative humidity.		
		Plant.	Atmom-eter.	Air.	Plant.	Atmom-eter.	Air.	Plant.	Atmom-eter.	Air.
		° *C.*	° *C.*	° *C.*	° *F.*	° *F.*	° *F.*	*p. ct.*	*p. ct.*	*p. ct.*
Mar.13.	5h17m p.m.	2.0	2.0	2.0	72	72	72	27	27	27
Mar.14.	6 19 a.m.	−3.5	−3.5	1.0	[1]56	[1]56	[2]56	24	24	58
	6 55 a.m.	0.5	0.5	0.5	59	59	57	38	38	38
	8 00 a.m.	2.0	2.3	2.8	64	64	63	35	35	38
	8 25 a.m.	3.0	3.0	3.0	65	65	65	36	36	36
	9 29 a.m.	1.5	1.3	−0.5	72	73	71	25	24	23
	9 36 a.m.	−0.5	−0.5	−0.5	73	73	73	21	21	21
	10 55 a.m.	−0.5	−0.3	−0.5	84	85	84	15	15	15
	11 00 a.m.	−0.5	−0.5	−0.5	84	84	84	15	15	15
	12 25 p.m.	−3.5	−3.0	1.3	86	86	86	11	11	16
	12 33 p.m.	1.5	1.5	1.5	86	86	86	16	16	16
	1 55 p.m.	−3.5	−3.5	−0.1	88	89	89	10	10	13
	2 04 p.m.	0.0	0.0	0.0	89	89	89	13	13	13
	3 05 p.m.	−2.0	−1.7	1.0	89	90	90	11	11	14
	3 22 p.m.	1.0	1.0	1.0	90	90	90	14	14	14
	4 23 p.m.	−2.5	−2.0	2.0	84	84	84	12	12	18

[1]50 min. [2]48 min.

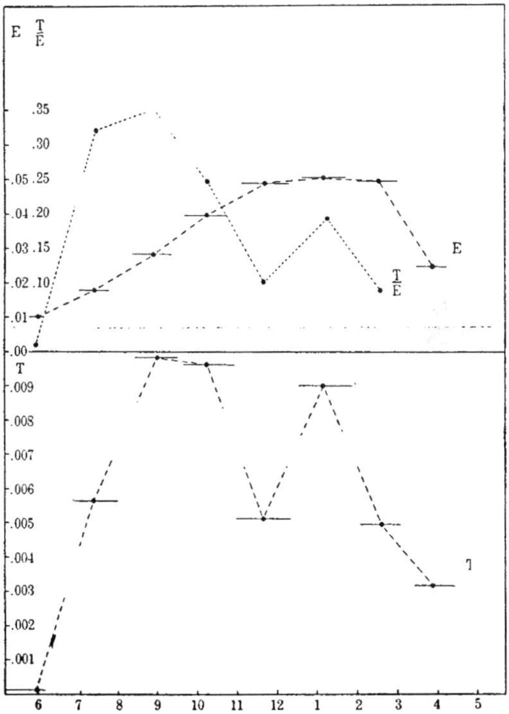

FIG. 8.—Graphs for transpiration of *d*, a branch in leaf from an adult tree. Exp. IV.

TABLE 5.—*Transpiration of a branch in leaf (e) of an adult tree. Experiment V. Date, March 29, 1911.*

Time.	T.	Time.	E.	T/E
7h 03m a.m. 8 04 a.m.	0 00382	7h 05m a.m. 7 58 a.m.	0.030	0.126
8 08 a.m. 9 14 a.m.	.00618	8 11 a.m. 9 10 a.m.	.037	.166
9 19 a.m. 10 19 a.m.	00835	9 21 a.m. 10 10 a.m.	.046	.181
10 24 a.m. 11 22 a.m.	.00398	10 26 a.m. 11 15 a.m.	.050	.079
11 36 a.m. 12 24 p.m.	.00636	11 38 a.m. 12 15 p.m.	.051	.124
12 39 p.m. 1 53 p.m.	.00545	12 42 p.m. 1 45 p.m.	.052	.105
2 11 p.m. 3 20 p.m.	.00480	.....	.....	.....
3 24 p.m. 4 28 p.m.	.00193	.....	.....	.....

Time.	Dew-point.			Temperature.			Relative humidity.		
	Plant.	Atmometer.	Air.	Plant.	Atmometer.	Air.	Plant.	Atmometer.	Air.
	° C.	° C.	° C.	° F.	° F.	° F.	p. ct.	p. ct.	p. ct.
7h 03m a.m.	−3.0	−3.0	−3.0	66	66	66	22	22	22
8 04 a.m.	−9.0	−8.5	−7.0	71	72	71	12	12	14
8 08 a.m.	−7.0	−7.0	−7.0	72	72	72	14	13	14
9 14 a.m.	0.5	0.2	−2.0	82	82	83	19	18	14
9 19 a.m.	−2.0	−2.0	−2.0	83	83	83	14	14	14
10 19 a.m.	0.0	0.5	−2.0	88	89	88	14	14	14
10 24 a.m.	−2.0	−2 0	−2.0	88	88	88	11	11	11
11 22 a.m.	0.0	0.3	−4.0	90	90	89	13	14	9
11 36 a.m.	−4.5	−4.5	−4.5	89	89	89	9	9	9
12 24 p.m.	−2.0	−1.6	−4.5	91	91	90	11	11	9
12 39 p.m.	−4.5	−4.5	−4.5	90	90	90	9	9	9
1 53 p.m.	−0.5	−1.6	−4.5	91	91	90	12	12	9
2 11 p.m.	−5.0	.....	−5.0	89	..	89	9	..	9
3 20 p.m.	−5.5	.....	−4.5	89	..	88	9	..	10
3 24 p.m.	−4.5	.....	−4.5	88	..	88	10	..	10
4 28 p.m.	−5.0	.....	−4.0	89	..	84	9	..	11

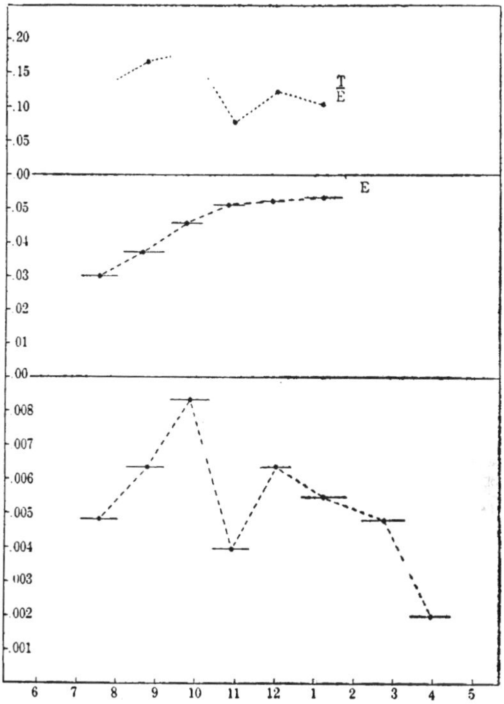

FIG. 9.—Graphs for transpiration of *e*, a branch in leaf from an adult tree.

EXPERIMENT VI.

The subject of this experiment was a potted plant which had been raised from seed in the green-house and then placed in the open for about 4 weeks before the beginning of this experiment. At this time the plant was about a year old. It is called No. 1 and appears in later experiments under the same number. The area was found to be 52.95 sq. cm., but since the leaves were not removed for measurement until April 24 a small amount of growth, which took place in the meantime, makes the amounts per square centimeter in this experiment slightly too small. The movement of the leaves differed from that of the trees, in that closing did not begin until after readings had been commenced. A record of the position of the leaves appears in this and subsequent experiments involving potted plants, in the form of two small lines placed below the corresponding readings on the curves. The angle of separation of these lines represents the angle of openness of the leaves. The relative humidity in the case of all potted plants was taken by a hygrograph record and not by the dew-point, as in the case of the bell-jar experiments. The atmometer used was No. 850 in the Livingston series and was found to be equivalent to 85 sq. cm. of free water surface. Plant and atmometer were placed in the open and brought into the laboratory for weighings. The wind velocity was high all the morning, remaining so without apparent change until 2 o'clock, when it increased to at least 25 miles an hour.

TABLE 6.—*Transpiration of potted plant No. 1. Experiment VI. Date, April 12, 1911.*

Time.	*T*	*E*	$\frac{T}{E}$	Temperature.	Relative humidity.
				° *F.*	*p. ct.*
7h 47m a.m.	....	......	.....	72	26
9 48 a.m.	0.00897	0.0361	0.248	78	22
10 47 a.m.	.00945	.0413	.228	83.5	20
11 48 a.m.	.00945	.0433	.218	84.5	11
12 48 a.m.	.00897	.0461	.194	89	9
1 47 a.m.	.00924	.0505	.183	88.5	8
2 47 a.m.	.00908	.0508	.178	86	8
3 48 a.m.	.00831	.0472	.176	80	9

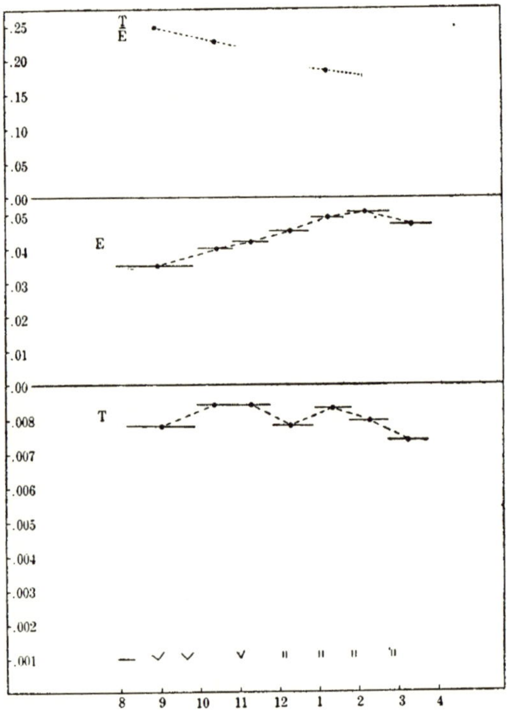

FIG. 10.—Graph for transpiration of potted plant No. 1. Exp. VI.

Experiment VII.

Subject, plant No. 1; area, 52.95 sq. cm. Atmometer No. 812, equivalent to 143 sq. cm. of free water surface. Situation, in the open, exposed to sun and wind, sun continuously bright all day. The wind movement was very slight up to 3 p. m., when the velocity increased markedly and continued high until the close of the experiment.

Table 7.—*Transpiration of potted plant No. 1. Experiment VII. Date, April 23, 1911.*

Time.	*T*	*E*	$\frac{T}{E}$	Temperature.	Relative humidity.
				° *F.*	*p. ct.*
8h 51m a.m.	......	......	.....	75	23
10 05 a.m.	0.0129	0.0300	0.430	79	21
11 18 a.m.	.0192	.0363	.526	86	19
12 10 p.m.	.0149	.0405	.368	88	18
1 36 p.m.	.0198	.0440	.450	89	18
2 22 p.m.	.0197	.0435	.452	90	17
3 23 p.m.	.0198	.0472	.419	88	17
4 24 p.m.	.0139	.0570	.244	87	17
5 24 p.m.	.0100	.0475	.210	86	17

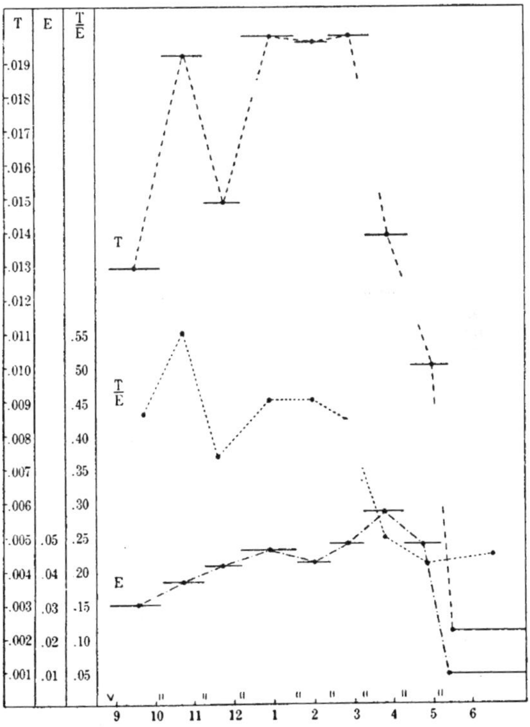

Fig. 11.—Graph for transpiration of potted plant No. 1. Exp. VII.

Experiment VIII.

Subject, plant No. 1; area, 52.95 sq. cm. Atmometer No. 812, equivalent to 143 sq. cm. of free water-surface. Situation, in the shade in a well-lighted room of the laboratory.

Table 8.—*Transpiration of potted plant No. 1, in the shade. Experiment VIII.*

Date.	Time.	T	E	$\frac{T}{E}$	Temperature.
					° *F.*
Apr. 23..	5h 24m p.m.	......	......	.....	..
Apr. 24..	9 59 a.m.	0.0022	0.0091	0.241	77
	11 03 a.m.	.0056	.0091	.615	78
	12 03 p.m.	.0068	.0091	.745	79
	1 14 p.m.	.0072	.0093	.774	80
	2 03 p.m.	.0081	.0099	.818	80
	3 17 p.m.	.0064	.0116	.551	83
	4 14 p.m.	.0060	.0122	.491	..
	4 44 p.m.	.0106	.0152	.697	..

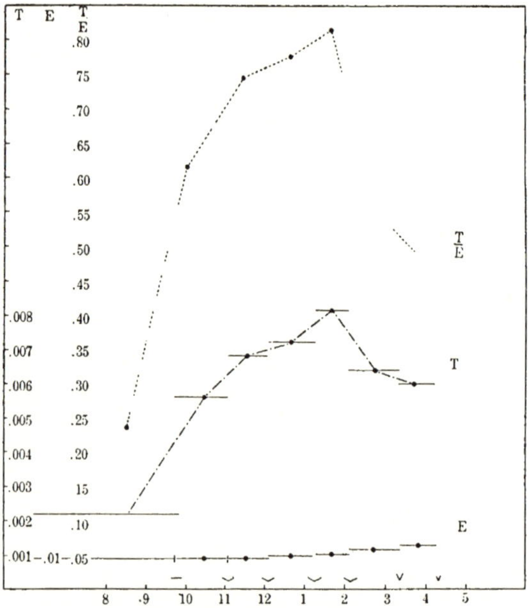

Fig. 12.—Graph for shade transpiration of potted plant No. 1. Exp. VIII.

Experiment IX.

Subject, four potted plants, Nos. 1, 2, 3, and 4 (see plate I*B*). Plants Nos. 2 and 3 had the same history as No. 1, while No. 4 was the only survival of some 30 which had been taken from their natural position in the soil and placed in pots several weeks before. No. 4 was a seedling which had germinated in the spring of 1910. It was in full leaf at the time of transplanting, but lost its leaves immediately and gained a new set after two weeks. The areas were as follows: No. 1, 68.31 sq. cm.; No. 2, 75.70 sq. cm.; No. 3, 101.67 sq. cm.; No. 4, 13.11 sq. cm. No. 1 had acquired a new set of leaves since April 25, when they were taken off for measurement.

TABLE 9.—*Transpiration of four potted plants. Experiment IX. Date, June 30, 1911.*

Time.	T_1	T_2	T_3	T_4	E_{10}	$\frac{T_1}{E_{10}}$	$\frac{T_2}{E_{10}}$	$\frac{T_3}{E_{10}}$	$\frac{T_4}{E_{10}}$	Temp.	Rel. hum.
										° *F.*	*p. ct.*
4h 17m a.m. 5 23 a.m.	0.00026	0 00125	0.00045	0.0001	0.0100	0.026	0.125	0.045	0.010	72 72	38 38
8 17 a.m.	.00758	.0074	.00728	.00495	.0363	.209	.204	.200	.136	78	37
10 12 a.m.	.0174	.0159	.0130	.00229	.0477	.367	.333	.273	.048	85	32
11 21 a.m.	.0193	.0168	.0130	.00526	.0502	.384	.334	.258	.105	89	31
12 18 p.m.	.0225	.0198	.0138	.00961	.0570	.395	.347	.242	.168	92	29
1 38 p.m.	.0212	.0196	.0152	.00850	.0640	.331	.306	.237	.132	97	28

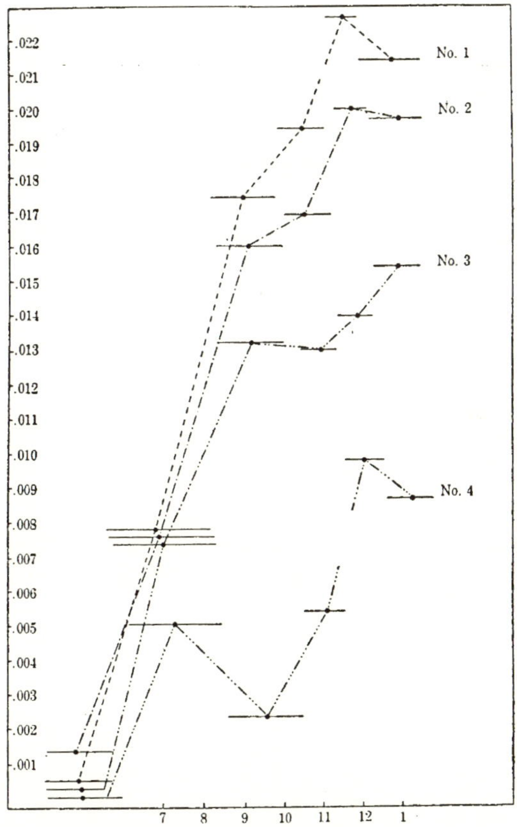

FIG. 13.—Graphs for actual transpiration of four potted plants. Exp. IX.

The atmometers used were No. 245, white, and a brown* one of the same type, which was not numbered in the series, but is called No. 10 for convenience. The white atmometer was accidentally destroyed before restandardization was made and therefore the readings for it are not comparable in actual amount to the brown cup nor to any other readings in the other experiments. The equivalent for the brown cup was 178 sq. cm. of free

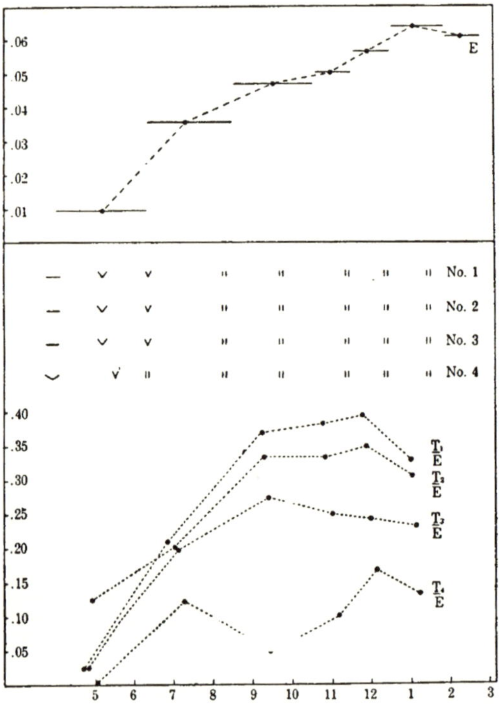

FIG. 14.—Graphs for relative transpiration of four potted plants, brown atmometer being used for evaporation rate. Exp. IX.

water surface. The situation was in the open with exposure to sun and wind. A heavy wind threw sand on the pots and thus spoiled the readings after $1^h 38^m$ p.m. The atmometer readings for 2^h 47^m p.m. show that the maximum evaporation for the day was between 12 noon and 1^h 30^m p. m. Air movement was confined to a very gentle breeze during the period covering the readings.

*Livingston, B. E., A radio-atmometer for measuring light intensities. Plant World, XIV, pp. 96–99, 1911.

EXPERIMENT X.

Subject, plants No. 1 and 2; areas, not obtained. Atmometers, No. 496 and brown cup No. 10. Situation, in the open, with exposure to sun and wind. On July 27 the leaflets were closed during the entire time of the

TABLE 10.—*Transpiration of potted plants under different conditions of soil moisture. Experiment X.*

Time.	T_1	T_2	E_{496}	E_{10}	$\frac{T_1}{E_{496}}$	$\frac{T_1}{E_{10}}$	$\frac{T_2}{E_{496}}$	$\frac{T_2}{E_{10}}$	$\frac{E_{10}}{E_{496}}$	Temperature.	Relative humidity.
July. 27.*										° *F.*	*p. ct.*
9h 33m a.m.	....	....	....	....	......	......	......	......	....	82	57
10 12 a.m.	2.51	1.26	3.40	4.10	0.738	0.612	0.370	0.307	1.20	86	47
11 09 a.m.	3.34	2.48	3.82	4.62	.874	.723	.649	.537	1 21	89	44
12 11 p.m.	3.34	2.48	4.64	5.12	.719	.652	.534	.484	1.10	91	43
1 04 p.m.	3.90	2.68	4.89	5.21	.797	.748	.548	.514	1.06	92	45
2 16 p.m.	3.36	2.24	4.30	4.65	.781	.722	.521	.481	1.08	92	45
3 16 p.m.	3.22	2.14	4.11	4.61	.783	.698	.521	.464	1.12	92	44
Aug. 5.†											
8 24 a.m.	....	....	....	....	......	......	......	......	....	89	36
9 30 a.m.	.09	.07	....	....	......	......	......	......	....	93	31
10 13 a.m.	.07	.09	4.90	6.22	.0143	.0112	.0183	.0144	....	96	32
11 17 a.m.	.05	.06	6.48	7.88	.0077	.0063	.0093	.0076	....	98	30
12 06 p.m.	.07	.06	....	....	......	......	......	......	....	101	22
1 15 p.m.	.10	.04	8.85	10.75	.0096	.0078	.0056	.0041	....	103	22
2 08 p.m.	.08	.08	....	....	......	......	......	......	....	103	22
3 09 p.m.	.07	.04	9.50	10.65	.0079	.0070	.0063	.0056	....	106	21
5 34 p.m.	....	....	7.79	9.16	......	......	......	......	....	...	..

* Soil moisture of No. 1=14.8 p. ct.
* Soil moisture of No. 2=15.1 p. ct.

† Soil moisture of No. 1=3.5 p ct.
† Soil moisture of No. 2=3.4 p. ct.

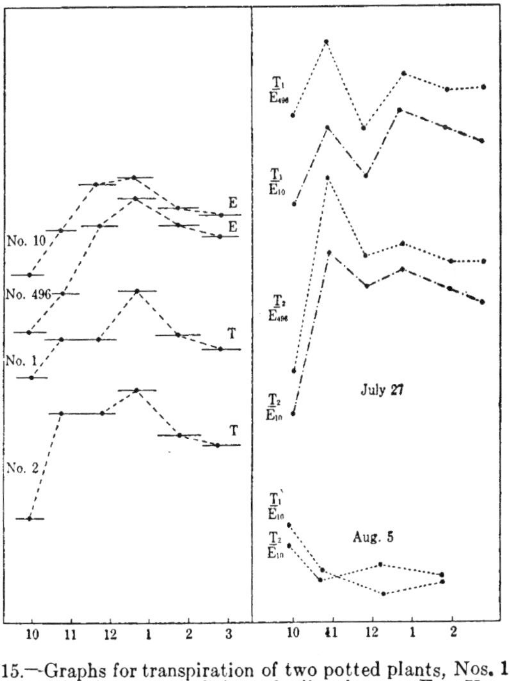

FIG. 15.—Graphs for transpiration of two potted plants, Nos. 1 and 2, under different conditions of soil moisture. Exp. X.

experiment, having closed 2 hours before the first reading. In this experiment, E represents the actual loss per hour from the atmometers multiplied by their proper coefficients. Thus, only the shape of the curves may be compared with the other experiments, and not thc amplitudes. The sun was uniformly bright until $2^h 30^m$ p. m., after which partial cloudiness began and continued for the rest of the day.

Immediately after the 3 o'clock reading the percentage of moisture of the soil in the pots was determined. Small glass vials were filled with samples of the soil by forcing them into the pots, two samples being taken from each pot, one slightly below the surface and the other about two-thirds of the way to the bottom. The vials were quickly stoppered and weighed, and later were dried to constant weight at about 110° C.

The two plants were then unsealed and left without water until August 5, when the hourly transpiration rate was again determined. During the interval no visible change in area had occurred, so that the relative transpiration rates for July 27 and August 5 are comparable, although the areas were not obtained. On August 5 the leaves on both plants were slightly curled and remained nearly closed continuously. After the close of the readings for August 5 the percentage of soil moisture was again determined. Both plants were given water at 6 p. m., and by 9 p. m. all the leaves of No. 1 had resumed their normal appearance and within three days the plant had begun to show growth. The leaves of No. 2 did not assume their normal appearance, but dropped off the next day; however, by August 9 new leaf-buds were visible on No. 2. Both plants continued to grow normally after this.

Experiments XI to XIII.

The subject of these experiments was an adult tree situated south of the laboratory. Three separate branches (*f*, area 28.72 sq. cm.; *g*, area 30.02 sq. cm., and *h*, area 31.82 sq. cm.) were used. The entire tree was in full leaf at the beginning of the experiment, but at the close some leaflets had begun to fall and several neighboring trees had lost most of their leaflets. As mentioned above, the methods used in these experiments differ from those used in the bell-jar experiments of 1911 in the following points: (1) the evaporating surface of the atmometer was parchment paper; (2) the atmometer was placed under the same bell-jar with the plants; (3) carbon dioxide suddenly released from high pressure was used to cool the dew-point apparatus.

TABLE 11.—*Transpiration of a branch in leaf (f) of an adult tree. Experiment XI. Date, September 14, 1912.*

Time.	T	E	$\frac{T}{E}$	Dew-point		Temperature.		Relative humidity.	
				Bell-jar.	Air.	Bell-jar.	Air.	Bell-jar.	Air.
				°C	°C.	°F.	°F.	p. ct.	p. ct.
7h 30m a.m. 8 20 a.m.	0 00223	0 0140	0.159	14.0 13.8	14.0 13.0	... 87	... 85	.. 36	.. 37
8 49 a.m. 9 46 a.m.	.00607	.0295	.205	12.5 12.3	12.5 10.8	87 92	87 90	34 28	34 27
10 00 a.m. 11 00 a.m.	.00342	.0422	.081	10.5 10.3	10.5 11.0	90 96	90 94	27 22	27 24
11 15 a.m. 12 16 p.m.	.00451	.0482	.093	11.0 10.2	11.0 9.0	94 98	94 96	24 22	24 20
12 28 p.m. 1 30 p.m.	.00503	.0548	.109	8.3 8.5	8.3 8.1	96 100	96 97	20 17	20 18
1 40 p.m. 2 43 p.m.	.00428	.0510	.084	8.1 8.0	8.1 3.4	97 103	97 100	18 16	18 12
2 55 p.m. 4 00 p.m.	.00290	.0413	.070	3.0 3.1	3.0 1.7	100 102	100 101	12 11	12 10
4 15 p.m. 6 00 p.m.	.00103	.009	.110	1.5 2.0	1.5 3.0	101 95	101 93	10 13	10 15

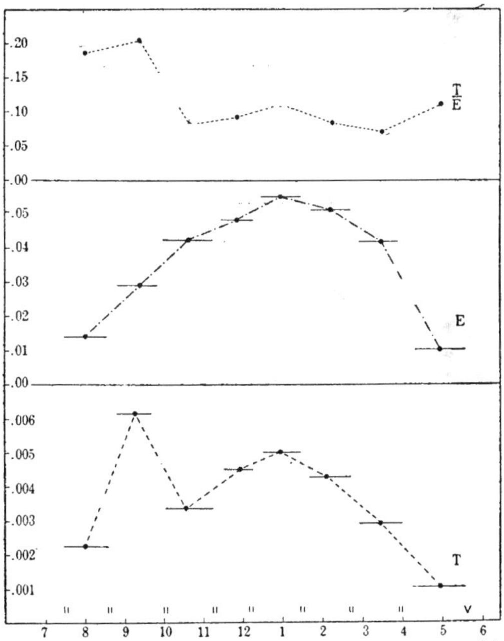

FIG. 16.—Graphs for transpiration of a branch in leaf (f) from an adult tree. Exp. XI.

TABLE 12.—*Transpiration of a branch in leaf (g) of an adult tree. Experiment XII. Date, September 16, 1912.*

Time.	T	E	$\frac{T}{E}$	Dew-point.		Temperature.		Relative humidity.	
				Bell-jar.	Air.	Bell-jar.	Air.	Bell-jar.	Air.
				° C.	° C.	° F.	° F.	p. ct.	p. ct.
6h 15m a.m. 7 15 a.m.	0 00150	0 0130	0.115	3.7 3.5	3.7 2.5	... 78	... 78	.. 24	.. 21
7 30 a.m. 8 36 a.m.	00492	.0298	.164	2.5 2.3	2.5 0.7	79 85	79 82	22 17	22 17
8 45 a.m. 9 47 a.m.	.00789	.0470	.168	0.5 0.0	0.5 −1.0	83 93	83 91	17 12	17 12
9 55 a.m. 11 00 a.m.	.00311	.0573	.055	−1.2 0.0	−1.2 −4.5	91 95	91 94	12 11	12 8
11 10 a.m. 12 15 p.m.	.00655	.0652	.100	−4.5 −3.0	−4.5 −4.5	94 99	94 97	8 8	8 7
12 26 p.m. 1 28 p.m.	.00566	.0690	.082	−4.6 −4.0	−4.6 −5.0	98 102	98 100	7 7	7 7 (?)
1 35 p.m. 2 40 p.m.	.00527	.0679	.076	−5.0 −4.7	−5.0 −4.5	101 104	101 102	7 7 (?)	7 7 (?)
2 50 p.m. 3 55 p.m.	.00445	.0588	.077	−4.5 −4.5	−4.5 −3.2	102 100	102 98	7 (?) 7	7 (?) 7
4 05 p.m. 5 06 p.m.	.00270	.0450	.060	−3.0 −2.0	−3.0 −1.5	97 95	97 94	7 10	7 9
5 15 p.m. 6 35 p.m.	.00100	.0066	.015	−1.5 −2.0	−1.5 2.0	92 80	92 80	10 15	10 20

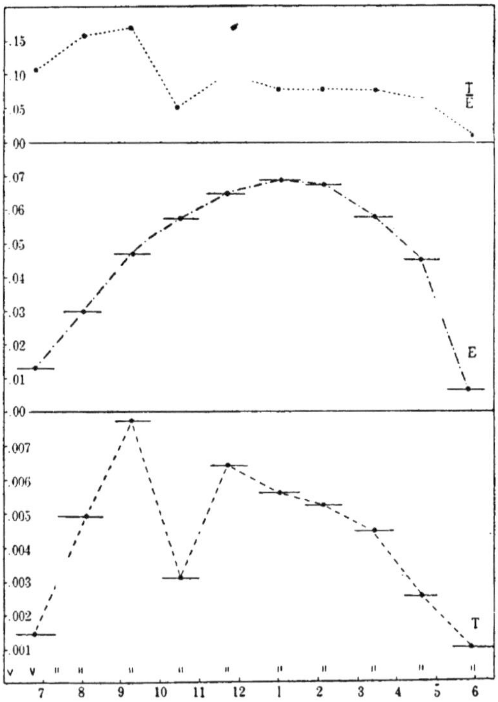

FIG. 17.—Graphs for transpiration of a branch in leaf (*g*) from an adult tree. Exp. XII.

Table 13.—*Transpiration of a branch in leaf (h) of an adult tree. Experiment XIII. Date, September 16, 1912.*

Time.	T	E	$\frac{T}{E}$	Dew-point.		Temperature.		Relative humidity.	
				Bell-jar.	Air.	Bell-jar.	Air.	Bell-jar.	Air.
				°*C.*	°*C.*	°*F.*	°*F.*	*p. ct.*	*p. ct.*
6h 32m a.m. 7 42 a.m.	0 00521	0 0245	0 212	−3.0 −2.5	−3.0 −2.0	. . . 80	. . . 78	17 15	. . 16
7 55 a.m. 8 45 a.m.	.00770	.0333	.231	−2.0 −2.5	−2.0 −1.9	78 86	78 83	16 12	16 14
8 55 a.m. 10 00 a.m.	.00692	.0410	.169	−1.0 −2.0	−1.0 −3.0	83 95	83 93	14 10	14 9
10 10 a.m. 11 13 a.m.	.00402	.0578	.069	−3.0 −2.7	−3.0 −3.8	93 98	93 95	9 8	9 8
11 25 a.m. 12 25 p.m.	.00623	.0608	.102	−4.0 −3.6	−4.0 −4.0	95 100	95 98	8 7	8 7
12 35 p.m. 1 39 p.m.	00598	.0612	.097	−4 0 −3.8	−4.0 −6.5	98 103	98 100	7 9	7 7 (?)
1 50 p.m. 2 53 p.m.	.00485	.0581	.083	−6.5 −3.8	−6.5 −4.0	100 103	100 101	7 (?) 9	7 (?) 7
3 05 p.m. 4 05 p.m.	.00347	.0436	.079	−4.0 −3.9	−4.0 −3.0	101 98	101 95	7 (?) 7	7 8
4 15 p.m. 5 20 p.m.	.00097	.0120	.080	−3.0 −2.0	−3.0 1.0	95 83	95 84	8 14	8 16

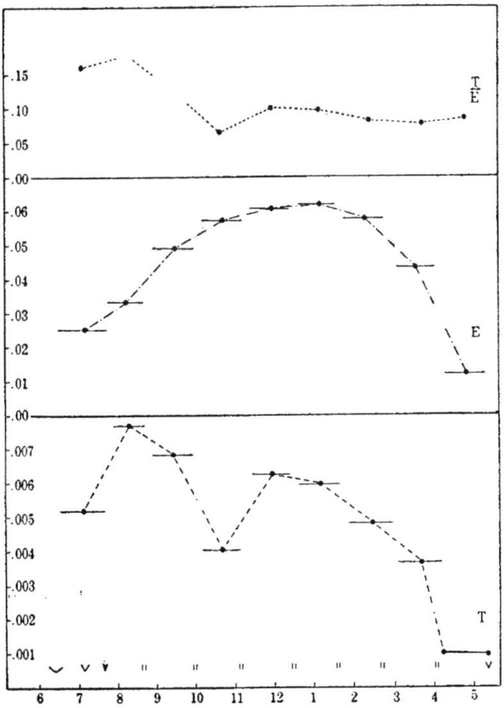

Fig. 18.—Graphs for transpiration of (*h*) a branch in leaf from an adult tree. Exp. XIII.

OBSERVATIONS ON TRANSPIRATION EXPERIMENTS.

1. The 18 relative transpiration curves taken in the sun all show a maximum earlier in the day than the time of maximum evaporation.

2. All of the relative transpiration curves, except those from experiments I, II, VI, and a part of IX, show an abrupt drop after the maximum, followed by a rise which appears sometimes an hour earlier than the time of maximum evaporation and sometimes coincident with it, but never later.

1 A. An early maximum in actual transpiration corresponding to that of the relative transpiration appears in all of the plants growing in the open, except those in experiments I and II. In the case of potted plants this maximum behaves somewhat irregularly. In experiment VI it is not pronounced, in VII it is lower than a second maximum, in IX it fails to appear in the case of three hot-house-grown plants and appears as a secondary maximum in the case of the transplanted seedling.

2 A. All curves of actual transpiration of plants growing in the open, with the exception of I and II, show a drop and rise in the morning. As in the case of the relative curves, the potted plants differ among themselves. Experiment VI shows the drop and rise to a small extent; IX shows a flattening of the slope in the morning followed by an increase in the angle of slope in the three hot-house plants and a drop followed by a rise for the transplanted seedling; X shows straight lines followed by a rise at the time of the drop in the relative transpiration.

3. It will be seen from table 14 that leafless seedlings and leafless branches of adult trees give the same average for maximum relative transpiration for the day, and that this average is lower than the average maximum for branches in leaf, and this is in turn much lower than the average maximum for potted plants. The one seedling which was transplanted from the open shows a maximum somewhat under the maximum average of branches in leaf, but is higher than two of the individual maxima of which the average is made up.

4. The differences in anatomical features shown in the sections in figure 1 indicate that the same relation exists between the water-losing power in leaves as compared to stems and in green-house plants as compared to those grown in the open, as is shown in the preceding paragraph.

5. Experiment X shows that a change of soil moisture in one potted plant from 15.1 per cent to 3.5 per cent, *i. e.*, a lowering of 76.8 per cent, was accompanied by a lowering of 98.4 per cent of the original amount of relative transpiration; in a second plant a change of the soil moisture from 14.5 to 3.4 per cent, *i. e.*, a lowering of 76.6 per cent, was accompanied by a lowering of 97.2 per cent of the original relative transpiration. In only one case were valid readings of soil moisture, at a depth of 30 cm., taken at times corresponding to transpiration readings from trees, and this shows that a change in soil moisture from 14.0 per cent on March 27 to 32.0 per cent on August 22, *i. e.*, an increase of 128 per cent, was accompanied by a change in relative transpiration from 0.181 on March 27 to 0.302 on August 22, *i. e.*, an increase of 72.3 per cent.

DISCUSSION OF TRANSPIRATION EXPERIMENTS.

Livingston* found a similar early maximum for relative transpiration in the case of potted plants of four species, but in general he did not find an early maximum in the actual transpiration, nor does he mention the rise succeeding the drop. It is possible that his readings were taken at too long intervals to bring out the phenomena. In the above experiments, Nos. I and II are mentioned as exceptions, but in these cases readings were taken at too great intervals for the drop and rise to appear if it existed. It may be noted that their afternoon curves for relative transpiration approximate straight lines. Although the drop and rise do not appear in plants Nos. 1 and 2 used in experiment IX, there is a corresponding decrease in the angle of the slopes of both the actual and the relative transpiration-curves which is followed by an abrupt angular increase. Plant No. 3 of this experiment shows the drop in relative transpiration, but no rise up to the time the experiment had to be discontinued, while No. 4 shows both drop and rise. In the case of experiment VI the closing of the leaves took place at such a time as to interfere with conclusions regarding the drop and rise. The closing of the leaves undoubtedly cuts down transpiration from the dorsal surfaces, since the air in the narrow space between the leaves would quickly rise in humidity. The second hour during which the leaves were closed showed a slight rise in actual transpiration over the first hour, but this made a very small change in the angle of the relative transpiration curve and might easily be due to an increase in transpiration from the ventral surfaces resulting from an increased vapor pressure within the leaf which would be caused by the lessening of evaporation from the dorsal surface. Thus it seems necessary to omit experiment VI altogether from the discussion of the occurrence of the drop and rise mentioned. In all the other experiments the closure of leaves took place too early in the day to influence the drop and rise. It will be noted that in the cases of trees, where the drop and rise is most pronounced, the leaves were closed by 5 a. m.

There is, then, in all of the above experiments, no clear negative evidence for the existence of a drop and rise; and from the experiments the generalization may be made that in trees, leafless and in leaf, a distinct drop and subsequent rise occur in both the actual and relative transpiration, in the morning before the time of maximum evaporation for the day; and that potted plants show a tendency towards the same phenomena, but in some cases the drop appears only as a straight-line "jog" in the relative transpiration curve.

An inspection of the values summarized in table 14 shows the following. The relative transpiration maxima which occur immediately preceding the drop in the relative transpiration curves vary so much that it can not be said that any relation exists between the amount of relative transpiration and the occurrence of the drop. However, the values for the actual transpiration at this period do show a tendency towards a constant value for the different forms used, so that it may be said that approximately 0.0044

*Livingston, B. E., Carn. Inst. Wash. Pub. 50, p. 63.

gram per square centimeter represents the average maximum for leafless forms, 0.0079 gram for adult trees in leaf, and 0.0164 gram for green-house-raised plants. Experiment VI was omitted from the series because of the error in area which has already been mentioned; so also was VIII, because it represents readings in the shade and does not show the drop.

TABLE 14.—*Summary of some of the data from Experiments I to XIII.*

Subject.	Maximum preceding drop in $\frac{T}{E}$			Minimum during drop in $\frac{T}{E}$			Maximum succeeding drop in $\frac{T}{E}$			Maximum for day.	
	$\frac{T}{E}$	T	E	$\frac{T}{E}$	T	E	$\frac{T}{E}$	T	E	$\frac{T}{E}$	T
Leafless seedling:											
Ex. I	0.213	0.0040	0.019	0.061	0.0056	0.052	0.072	0.0028	0.040	0.213	0.0056
Ex. II	.136	.0045	.033	......	......	......	......	......	......	.136	.0052
Ex. II	.120	.0041	.034	......	......	......	......	......	......	.120	.0048
Average	.156	.0042	......	......	......	......	......	......	......	.156	.0051
Leafless branch of tree:											
Ex. III	.151	.0057	.038	.068	.0034	.049	.075	.0038	.050	.151	.0060
Ex. III	.158	.0035	.038	.069	.0031	.044	.078	.0038	.049	.158	.0043
Average	.155	.0046	......	.069	.0033	......	.077	.0038	......	.155	.0052
Branch of tree in leaf:											
Ex. IV	.353	.0098	.028	.104	.0051	.049	.174	.0090	.052	.353	.0098
Ex. V	.181	.0083	.046	.079	0040	.050	.124	.0064	.051	.181	.0083
Ex. XI	.205	.0061	.029	.081	.0034	.042	.109	.0050	.055	.205	.0061
Ex. XII	.168	.0079	.047	.055	.0031	.057	.100	.0065	.065	.168	.0079
Ex. XIII	.231	.0077	.033	.069	.0040	.058	.102	.0062	.061	231	.0077
Average	.227	.0079	......	.078	.0036	.....	.122	.0066	......	.227	0079
Potted plant No. 1 green-house-raised, Ex. VII	.526	.0192	.036	.368	.0149	.040	.452	.0198	.044	.526	.0198
Potted plant No. 1, green-house-raised, Ex. IX	.426	.0174	.047	.445	.0193	.050	.459	.0225	.049	.459	.0225
Potted plant No. 2, green-house-raised, Ex. IX	.387	.0159	.041	.387	.0168	.050	.404	.0198	.049	.404	.0198
Potted plant No. 3, green-house-raised, Ex. IX	.317	.0130	.041	none	.0130	.050	none	.0152	.049	.317	0152
Average	.414	.0164	......	......	.0160	......	......	......	......	.428	.0192
Potted plant No. 4, transplanted from open, Ex. IX	.158	.0049	.031	.056	.0023	.047	.195	.0096	.049	.195	.0096

The maxima for experiment IX have been reduced to white-atmometer maxima by multiplying them by 1.162, a correction factor which was found by averaging the results obtained by dividing the brown-atmometer readings by the corresponding white-atmometer readings on July 27 and on August 5. (See experiment X.)

Considering only those experiments on plants growing under natural conditions which show a definite drop and rise in both the actual and relative transpiration, it appears that the maximum of actual transpiration reached after the drop is in five cases lower than the one before the drop, and in two cases the later maximum is about the same as the early one, while in no case is the second one higher than the first. It therefore appears that, for a certain form under similar conditions, there is a maximum water-loss beyond which the plant does not go and that this is frequently reached without a previous decrease in the relative rate. In the case of green-house-raised plants, where, as has been pointed out, the drop is replaced by a lessening of the increase of relative rate and is followed by an abrupt increase in relative rate, it appears that while there is a tendency toward a constant value for the actual transpiration immediately preceding diminution in the increase of the relative rate, yet it is not so marked as in the case of plants *in situ*. In three out of four cases the later maximum is higher than the first. In these cases, though, the relative transpiration rate again decreases before the time of maximum transpiration rate. In all of the four cases mentioned the soil moisture in the pots was probably higher than that in the ground, and since experiment X throws some light on the behavior of maxima in both relative and actual rates under different conditions of soil moisture, it is not possible from the potted plants to draw conclusions strictly comparable with those drawn from the plants *in situ*.

It seemed probable that this drop and rise have some definite internal cause, and with the idea of obtaining some light on the cause, measurements were taken of the following factors: (1) stomatal openings; (2) water content of leaves, stems, and twigs; (3) temperature of leaves.

STUDIES INVOLVING FACTORS CORRELATED WITH TRANSPIRATION BEHAVIOR.

DAILY COURSE OF STOMATAL MOVEMENT.

METHODS.

For the investigation of stomatal behavior in *Parkinsonia* the absolute alcohol method used by Lloyd* was tried and found to be satisfactory in the case of leaves of potted plants and of very young plants growing in the open; but it could not be used on the leaves or twigs of adult trees, since in the case of the leaves much of the mesophyll adhered to the peeled epidermis and in the case of the twigs the epidermis came off without the guard cells, which are sunken in deep pits. An effort was made to find some method by which the leaf as a whole could be used. Picric acid was dried at about 40° C., cooled, and put into cedar oil until a saturated solution with picric acid in suspension was made. Whole leaves and parts of leaves were put into this mixture and left there until they were cleared sufficiently to be seen by reflected electric light through a microscope. Picric acid was used because of its known properties as a "fixative." Since an aqueous solution could not be used without causing a disturbance of the water content of the cells, cedar oil was selected as a solvent, on account of its ability to penetrate and "clear" fresh plant tissues. The theory of the action of this mixture on the tissues is pure conjecture, but since it was the theory that suggested the use of the materials it is given here for what it may be worth. The picric acid is carried by the cedar oil into the tissues of the leaf, and as soon as it meets the water in the cell walls it goes into aqueous solution and then acts upon the protoplasmic lining within, making it impermeable to water and thus leaving the form of the guard cells undisturbed.

The only test for this method is of course the checking of results secured with those obtained by some other method. Since the leaves of potted *Parkinsonia* plants lend themselves to Lloyd's absolute alcohol method, they were used to test the efficiency of the picric acid method, readings being taken by the two methods simultaneously. An extra check was attempted by the use of the direct light method used also by Lloyd (not yet published), but the leaves did not lend themselves to this method.

Experiment for Testing the Picric Acid Method.

Two species of *Tradescantia*, *Cissus laciniata*, and a hot-house-grown *Parkinsonia microphylla* were used for the test. Lloyd's absolute alcohol method worked satisfactorily on all of the plants, but neither the light method nor the picric acid method could be used on either *Tradescantia*. The picric acid method worked satisfactorily on both the *Cissus* and the *Parkinsonia*. The numbers given in the tables represent averages of 20 readings each, which were obtained as follows: The entire preparation was examined, the probable average size estimated, and 18 individuals near this size were measured; then one near each extreme was measured and included.

*Lloyd, F., Physiology of stomata, Carn. Inst. Wash. Pub. 82.

In most instances the majority of the stomata were about the average size. A test of the method of selection was made by taking a leaf at random and measuring all the stomata included in it. The average was found to be practically the same as the one obtained by the first method. In the opinion of the observer the widths given represent very little error, but where the openings were small the lengths could not always be determined accurately. Every measurement was made to include equal numbers of stomata on the two sides of the leaves. The necessity for the inclusion of the stomata of both sides of the leaves in measurements of this kind has been pointed out by Livingston and Estabrook.*

Table 15 contains the results reduced to microns, including the length (l) of the openings between the guard cells, the width (w) of the opening, and the square root of their products ($\sqrt{lw}$).†

A comparison of the curves shows that the method under discussion gives approximately the same results as Lloyd's absolute alcohol method. In one respect this method seems to have an advantage over the absolute alcohol method, namely, that the guard cells are not subjected to the danger of loss of water from evaporation during the interval of transfer from the leaf to the alcohol, a factor which might be serious in the dry atmosphere of Arizona.

TABLE 15.—*A comparison of stomatal measurements taken by the absolute alcohol method and the picric acid method. Cissus and Parkinsonia, November 17, 1912.*

	Absolute alcohol.			Picric acid in cedar oil.			Light.		
	l	w	$\sqrt{lw}$	l	w	$\sqrt{lw}$	l	w	$\sqrt{lw}$
Cissus.									
8h 00m a.m.	7.1	1.4	3.1	9.1	1.4	3.5	....	...	...
9 30 a.m.	9.1	3.9	5.9	9.9	3.3	5.8	....	...	...
12 00 noon	9.9	3.2	5.6	9.9	2.4	5.0	....	...	...
2 30 p.m.	8.3	2.3	4.3	9.5	1.7	4.0	....	...	...
5 30 p.m.	7.6	1.1	2.9	10.6	1.1	3.2	....	...	...
9 00 p.m.	7.6	.9	2.6	9.1	1.4	3.5	....	...	...
Parkinsonia.									
8h 00m a.m.	9.1	1.5	3.7	9.1	1.5	3.7	10.6	2.3	4.9
10 30 a.m.	9.1	2.7	4.9	9.1	2.8	5.0	10.6	3.0	5.6
12 30 a.m.	7.6	1.1	2.9	7.6	.8	2.5	....	...	...
2 30 a.m.	7.6	.8	2.4	6.1	.9	1.5	....	...	...
5 30 a.m.	8.3	2.0	4.0	9.1	2.4	4.6	....	...	...
After 5 hours in dark room	7.6	.7	2.6	8.3	1.5	3.5	....	...	...
After 24-hours in dark room	Shut.	0	Shut.	Shut.	0	Shut.	....	...	...

EXPERIMENTATION.

EXPERIMENT XIV.

During the progress of experiment V (see page 22) leaf samples for stomatal measurements were taken as near the middle of each transpiration period as the manipulation of the experiment would permit and treated with picric acid in cedar oil, as described above. The values of T/E are repeated from page 23 and are plotted in fig. 20 with the values for the square roots of the products of the stomatal diameters.

*Livingston, B. E., and Estabrook, A. H., Observations on the degree of stomatal movement in certain plants, Bull. Torr. Club, xxxix, 15–22, 1912.

†Browne and Escomb, Static diffusion of gases, Phil. Trans. Roy. Soc. Lond., B, vol. 192, pp. 270–292, 1900.

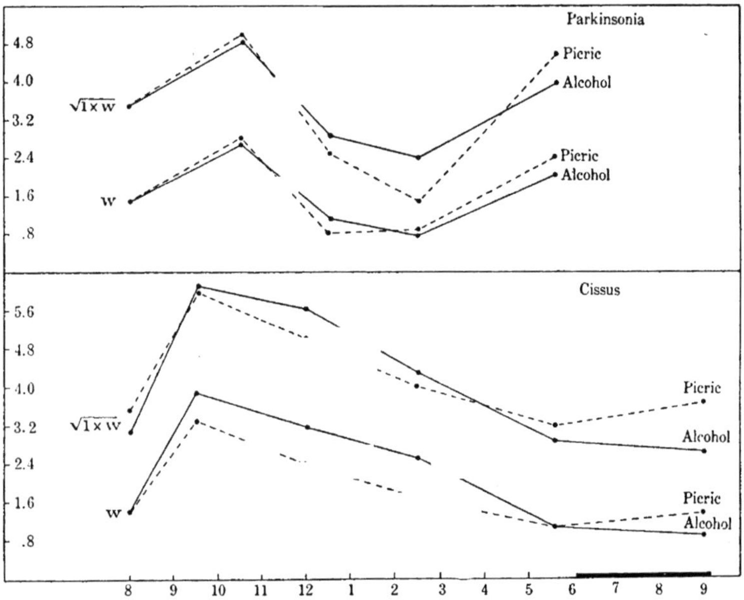

FIG. 19.—Comparison of picric-acid method for stomatal measurement with the absolute alcohol method.

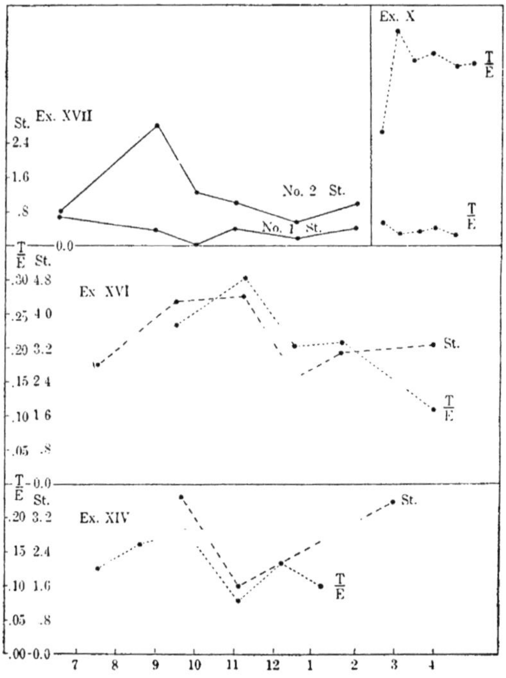

FIG. 20.—Graphs for stomatal openings and relative transpiration. Exps. XIV, XVI, XVII.

TABLE 16.—*Relative transpiration and stomatal openings. Experiment XIV. Date, March 29, 1911.*

Time.	l	w	$\sqrt{lw}$	$\frac{T}{E}$
9h 40m a.m. shade....	8.8	1.4	3.5	0.181
9h 40m a.m. sun......	8.8	1.6	3.7	
11 05 a.m............	4.4	.6	1.6	.079
12 10 p.m............	6.0	.8	2.2	.124
3 00 p.m............	8.8	1.5	3.6	.....

EXPERIMENT XV.

Leaves and tangential surface sections of young twigs from tree *A* were taken simultaneously with leaves from plant No. 1, subjected to the picric acid treatment, and stomatal measurements were made. As a control, stomatal measurements were also made on No. 1 by Lloyd's method, through the 1 o'clock reading. The potted plant was placed in the open, about 50 feet distant from the tree. It will be seen that the agreement of the alcohol method with the picric acid method here is not as good as in the test experiments; however, the lack of agreement is one of actual size only, for the two curves have practically the same shape. The guard cells of the stomata of the tree were always smaller than those of the potted plant. The stomata of the twigs very evidently are not movable to any extent. A few days after this experiment measurements of the stomata of twigs were tried and no evidence of any motion of the stomata was gained from some 20 trials.

TABLE 17.—*Stomatal measurements from leaves and twigs of an adult tree and from leaves of a potted plant. Experiment XV. Date, August 11, 1911.*

Time.	No. 1 leaves.						*A* leaves, picric acid.			*A* twigs, picric acid.		
	Picric acid.			Absolute alcohol.								
	l	w	$\sqrt{lw}$	l	w	$\sqrt{lw}$	l	w	$\sqrt{lw}$	l	w	$\sqrt{lw}$
4h 30m a.m.	4.2	0.6	1.4	4.2	0.6	1.6	5.6	1.0	2.4	.....	......	..
6 00 a.m.	11.6	4.6	7.2	8.8	...	6.0	6.4	1.3	2.9	.....	0	0
9 00 a.m	10.8	3.8	6.4	8.8	2.5	4.7	6.4	3.4	4.7	[1] 4.8(2)	[1] .4(2)	0
11 00 a.m.	8.8	2.4	4.5	8.8	1.5	3.6	7.6	1.1	2.9	([2])	([2])	([2])
1 00 p.m.	12.0	2.1	5.0	7.6	1.6	3.4	7.6	.9	2.7	([2])	([2])	([2])
2 00 p.m.	10.8	3.5	6.1	...	...	...	6.4	1.5	3.1	[1] 4.8(4)	[1] .4(4)	..
4 00 p.m.	10.8	4.4	6.9	...	...	...	7.2	2.5	4.2	([2])	([2])	([2])
10 00 p.m.	6.4	1.0	2.7	...	...	...	6.4	.6	1.9	[3] 12.0	[3] .8	..

[1]All the rest shut. [2]All shut. [3]Average of 6; all the rest shut.

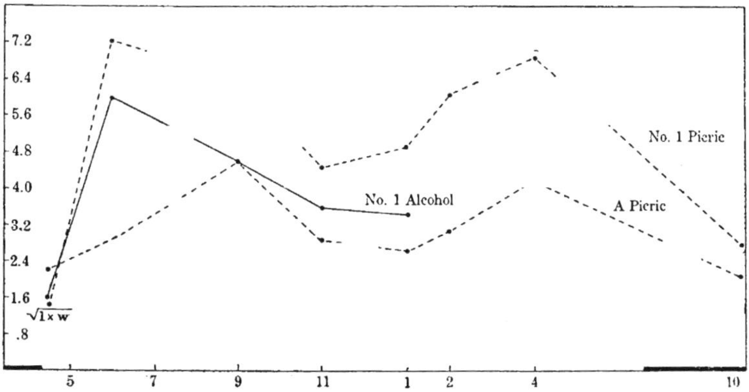

FIG. 21.—Stomatal measurements from leaves of potted plant and from adult tree by the picric-acid method, with absolute-alcohol method check on potted plant. Exp. XV.

Experiment XVI.

The subject of this experiment was an adult tree, called *B*, growing about 20 feet from tree *A*. Transpiration rate was determined by the calcium chloride method described on page 10, and the evaporation was determined by the use of atmometer of type 2 (*C*, fig. 2) under a separate bell-jar. The area of the branch used was not determined, so that the *T* here represents actual losses and not losses per unit area. Figure 20 gives the values for $\sqrt{lw}$ plotted with the values for T/E.

Table 18.—*Relative transpiration and stomatal openings from an adult tree. Experiment XVI. Date, August 23, 1911.*

Time	l	w	$\sqrt{lw}$	E	$\frac{T}{E}$
7h 30m a.m.	8.2	.9	2.8	.....	.....
9 30 a.m. shade	8.0	1.4	3.3	0.032	0.235
9 30 a.m. sun	8.0	2.3	4.3		
11 15 a.m.	8.0	2.4	4.4	.040	.302
12 30 p.m.	8.0	.8	2.5	.050	.205
1 40 p.m.	8.0	1.2	3.1	.052	.215
4 00 p.m.	8.0	1.4	3.3	.048	.108

Experiment XVII.

Method, Lloyd's for measurement of stomata.

Subjects, plants Nos. 1 and 2, and atmometer No. 10, brown. Situation, in the open, exposed to sun and wind. Conditions, soil-moisture conditions were made to correspond as nearly as possible to those of July 27 and August 5. No. 1 had been without water 5 days, while No. 2 had received the usual amount. The leaves of No. 1 had the same curled appearance as on August 5, while No. 2 appeared in all respects normal, as on July 27. At the close of the experiment, No. 1 responded to water as on August 5. The percentage of soil moisture was determined as before and appears in the tables. In this experiment only the widths of the openings of the stomata are given, since the curve of widths has been found to correspond in shape to the curve of $\sqrt{lw}$. All measurements were taken in the bright sunlight unless otherwise stated. At 6h 30m a. m. and 12h 30m p. m. measurements were also made in the shade and appear on the curves as isolated dots above the corresponding hours. To the left of the curve appear again, reduced one-half, the relative transpiration rates for plants Nos. 1 and 2 under similar conditions of soil moisture. So far as can be told from the evaporation readings for the 3 days, the evaporation conditions were similar.

Table 19.—*Comparison of stomatal openings of potted plants under different conditions of soil moisture. Experiment XVII. Date, August 20, 1911.*

Time.	No. 1, *w*.*	No. 2, *w*.†	*E* No. 10.
6h 30m a.m. shade	1.4	...	....
6h 30m a.m. sun	.7	.8	....
9 00 a.m., sun	.4	2.8	5.23
10 00 a.m., sun	0	1.2	6.37
11 00 a.m., sun	.4	1.0	7.09
12 30 p.m. shade	1.1	3.8	8.21
12 30 p.m. sun	.2	.6	
[1]2 00 p.m.	.4	1.0	9.05

* Soil-moisture of No. 1=3.6 p. ct.
† Soil-moisture of No. 2=14.2 p. ct.
[1] Cloudiness prevented further readings.

DISCUSSION OF STOMATAL BEHAVIOR.

In all of the seven experiments given above, the stomata of *Parkinsonia microphylla* closed down and subsequently opened during daylight. Three of the closures took place on adult trees and four on potted plants. In all cases the closure began before the time of maximum evaporation for the day had been reached. In the two cases (experiments XIV and XVI) where transpiration and stomatal measurements were taken simultaneously from different parts of the same tree, the curves of transpiration run parallel with the curves of the linear dimensions of the stomatal openings during the forenoon, but in the afternoon depart and run in opposite directions. These two experiments present some evidence that there is a connection between stomatal opening and transpiration rate and that the connection is not operative in the afternoon. The curves follow each other so closely during the morning hours that it seems impossible to neglect their evidence, even though the afternoon readings show evidence of the opposite sort. The fact that the time of minimum transpiration corresponds with the time of minimum stomatal opening in both experiments, and that this time is not the same in the two experiments can not be ignored. It is well to note here that between the dates of the two experiments the summer rains had come and the soil moisture at a depth of 30 cm. had been raised from 14 to 32.1 per cent, and consequently the deeper layers of soil must have had their moisture content raised proportionally. The trees had lost their leaves and gained new ones between the two experiments. While there is no convincing evidence of the direction of the cause and effect, a point might be made against the theory that the closure of stomata is due to light effect alone, for the difference in time of closure in the two experiments could hardly have been brought about by the difference of position of the sun, since the sun was about at the same distance from the summer solstice at the two times.

A reference to pages 19–29 shows that while the time of the morning minimum relative transpiration is fairly constant for a given tree on successive days, it is by no means constant in the case of potted plants. The stomatal openings may be compared with the relative transpiration when the readings are taken simultaneously from the same tree, but a reference to experiment IX will show the impossibility of making such hourly comparisons in the case of two potted plants unless the plants used can be shown to have parallel hourly curves of transpiration. Several attempts were made to get stomatal readings from one potted plant and transpiration from another which had been found to give a similar relative transpiration curve, but various interruptions of weather and accidents to the plants prevented good readings. The curves given above show that there is some connection between the relative transpiration and the size of the stomatal opening, but there is no evidence of the relation of cause and effect nor even of priority of time.

Experiment X, which was an attempt to investigate this relation in another way, brings out the fact that the relative transpiration of a plant of potted *Parkinsonia microphylla* in a soil of 3 per cent moisture is only one-sixtieth to one-hundredth the amount the same plant gives when its soil moisture is 14 per cent, and that this change takes place without loss of leaves. Now, experiment XVII shows that when the same plants, which experiment X shows behave similarly under similar conditions of soil-moisture, are run together under identical atmospheric conditions but under different conditions of soil-moisture, the stomatal diameters in the plant having a soil of 3 per cent moisture are about one-seventh of those from the plant having a soil of 14 per cent moisture. Evidently a lowering of the soil moisture means a lessening of the relative transpiration and also a shutting up of the stomata, but it will be noted that the transpiration rates are not proportional to the linear dimensions of the stomatal openings.

Lloyd's calculations for *Fouquieria splendens** show that the closure of stomata is not the cause of changes in relative transpiration rate. This is perhaps true of other plants, and with this in view the following hypothesis was framed as the starting-point for the work which follows. The morning drop is caused by a deficiency in water in the transpiring tissues, due in turn to the supply from the lower tissues not being equal to the demand of the evaporating power of the air. This deficiency of water in the leaves causes a closure of stomata simply because water is drawn from the guard cells in the same manner as from the other cells. The diminished size of the openings would cause a rise of humidity within the intercellular spaces and thus a checking of the evaporation rate within the leaf, even though the stomata did not close down enough to check the theoretical amount of vapor diffused through the openings. Thus the closure may act to further check transpiration, and meantime the ratio of outgo to intake has lessened and all the cells, including the guard cells, gain water and the latter open mechanically, after which transpiration goes on more rapidly, but does not reach the former high rate unless the ratio of outgo to intake reaches unity. However, in the case of young plants without leaves the early maximum and drop occur; and in the case of the leafless branch of an adult tree the drop is followed by a subsequent rise. It will be seen from table 17 that the stomata of twigs remain nearly closed all of the time, and this leads to the conclusion that the lessening of the relative transpiration rate before the maximum evaporation rate for the day has been reached may take place without a change in stomatal opening.

In experiment I the readings were not taken at frequent enough intervals to discover a rise after the drop, if it occurred. In experiment II a flattening of the curve occurs after the drop, but no distinct rise; while in experiment IV the rise after the drop is less in amount than in cases where plants in leaf were used; this may mean that the closure and subsequent reopening of

*Lloyd, F. E., The physiology of stomata, Carn. Inst. Wash. Pub. 82, pp. 136–137, 1908.

the stomata aided in the rise but are not necessary to it. So, then, a deficiency of water in the evaporating tissues would be the first cause of the drop in transpiration rate, while the evaporation rate is still increasing. If now the supply of water from below is unchanged, the smaller evaporation rate would allow a balance to be obtained when we should expect the relative rate to become a straight line for a time at least. It is hard to see how the gain could be sufficient to send the relative rate up again unless some other factor entered, either to check the rate of outgo or to increase the rate of intake from the lower parts. If Dixon's hypothesis* of the ascent of sap be accepted, it might be supposed that the deficiency of water in the terminal tissues causes a "pull" on the water columns, which enter the tensile state, and thus the intake is increased. As soon as the supply begins to catch up with the demand the tensile pull is lessened and the relative rate begins to go down again.

If the explanation is the latter one, based on Dixon's hypothesis, then there might be a relation between the moisture content of different parts of the tree which would throw light on the matter. Consequently a series of moisture content determinations was carried out as, follows.

DAILY COURSE OF WATER CONTENT OF LEAVES, TWIGS, AND STEMS.

EXPERIMENTATION.

Experiment XVIII.

Three adult trees, designated I, II, III, were selected in open sunny places, and at two-hour intervals small limbs were sawed off and samples for the determination of water-content were taken as follows: I_1, II_1, and III_1 were end twigs about 3 to 5 cm. long, with diameters from 0.1 to 0.2 cm.; I_2, II_2, and III_2 were taken about a meter from the end twigs and had diameters about 1.2 to 1.6 cm.; I_3, II_3, III_3, were taken 2 meters from the end twig and were from 3.5 to 3.9 cm. in diameter. The work of sawing and cutting was done very quickly, with the aid of an assistant, and the pieces were weighed immediately after the cuts were made. The samples were allowed to dry in the sun for about a week and then dried to constant weight in an oven at 95° to 100° C. Accidents which happened to the end twigs of II left too few good readings for the results to be plotted. For the sake of a comparison with the water-content of a tree which had lost no branches, samples were taken from a fourth tree at 2^h 30^m p. m. The two readings marked III_4 and IV_4 were taken about 4 meters from the tips, where the bark is brown and not green as in the other samples. In the sawing great care was taken to get branches widely separated on the tree, in order that the wounds might cause as little disturbance as possible. The evaporation rate was measured by exposing atmometer No. 11-1 in an open sunny place.

---

*Dixon and Joly, On the ascent of sap, Proc. Roy. Soc. Lond., vol. 57 B.

TABLE 20.—*Water content of stems and branches of adult trees. Experiment XVIII. Date, August 5, 1911.*

Time.	I_1	I_2	I_3	II_1	II_2	II_3	III_1	III_2	III_3	III_4	IV_1	IV_2	IV_3	IV_4	E_{11-1}
4h 30m a.m.	87	68	58	...	73	62	97	63	64	50	..	..	..	..	..
8 30 a.m.	85	66	68	87	68	61	93	60	61	..	..	..	..	..	..
10 30 a.m.	77	58	61	...	67	67	97	63	61	..	..	..	..	..	.037
12 30 p.m.	73	63	66	...	70	70	78	72	60	..	..	..	..	..	.046
2 30 p.m.	77	64	63	93	..	70	93	59	62	..	97	81	61	60	.061
5 00 p.m.	85	66	62	...	77	73	..	61	63	..	..	..	..	..	.055
9 30 p.m.	..	66	59	100	68	64	90	62	60	..	..	..	..	..	..

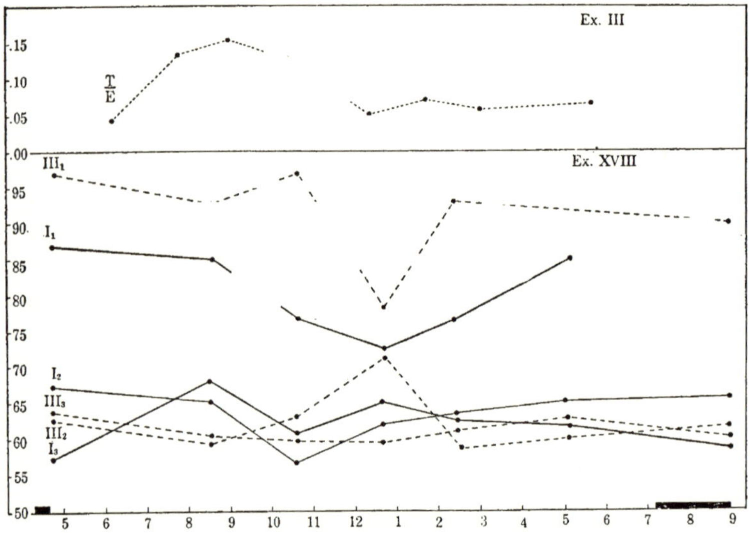

FIG. 22.—Graphs of water content measurements taken simultaneously from two adult trees at three different distances from the ends. Graph for relative transpiration of a leafless branch repeated at top of figure. Exp. XVIII.

EXPERIMENT XIX.

Subject, the tree from which branches *f*, *g*, and *h* were taken. (See experiments XI to XIII.) End twigs about 3 to 5 cm. long were cut off, the leaves quickly severed, and the leaves and twigs placed in separate air-tight shallow weighing-bottles. The bottles with their contents were weighed on good chemical balances to 0.001 gm., and the contents were then transferred to glass beakers and protected from dust, etc., by filter-paper covers. The beakers and their contents were then dried to constant weight in an electric oven at 72° to 75° C. By the use of the above precautions it was possible to obtain the water-content, of the small amount of material which it was necessary to use, accurately to 0.2 per cent. Calculations were made as before for percentage of dry weight of the material used. At noon and at 2 p. m. samples were taken from another tree about 30 yards distant. In table 21 these appear under the heading "Series II."

Since the manner of shading used in the transpiration method described above requires attention at least as often as every 10 minutes, it was not practicable to obtain both transpiration and water content readings on the same day, and in order to obtain the two sets of readings in comparable form the water content experiment was preceded and followed by transpiration readings, transpiration of branch *f* (see p. 31) being taken on September 14, water content on September 15, and transpiration of branch *g* on September 16. No samples were taken from the limb from which came the branches used for transpiration.

In order to obtain an approximation of the water content at the time corresponding to the transpiration readings, which it will be remembered were plotted to the middle of the time period, lines were dropped from the transpiration readings until they intersected the curves of water content

TABLE 21.—*Water content of leaves and their twigs. Experiment XIX.*

Date.	Time.	Series I.		Series II.	
		Leaves.	Twigs.	Leaves.	Twigs.
1912.		*p. ct.*	*p. ct.*	*p. ct.*	*p. ct.*
Sept. 14...	9h 00m p.m.	109.1	76.0	.....	....
Sept. 15...	7 40 a.m.	102.4	74.4	.....	....
	8 45 a.m.	97.3	78.7	.....	....
	9 45 a.m.	98.2	81.1	.....	....
	10 45 a.m.	98.2	78.7	.....	....
	12 00 m.	101.1	74.4	103.0	74.4
	2 00 p.m.	102.7	78.2	102 0	72.2
	4 00 p.m.	105.1	80.6	.....	....
	5 30 p.m.	103.3	78.1	.....	....
Sept. 16 ..	4 30 a.m.	109.4	78.2	.....	....

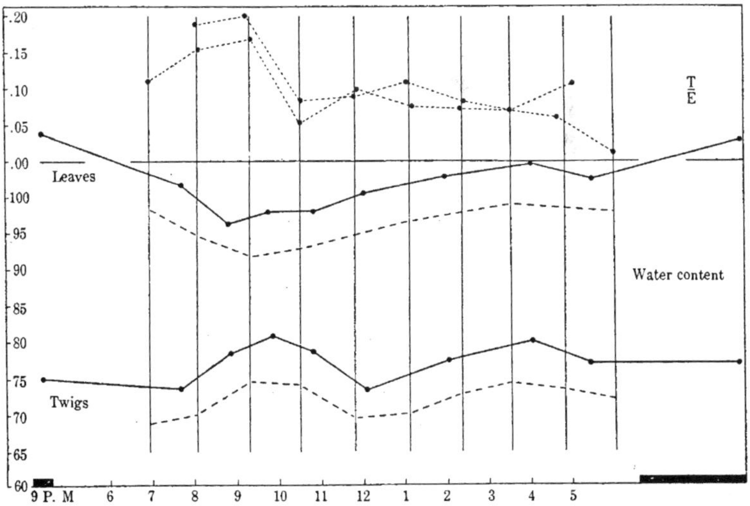

FIG. 23.—Graph for moisture content of twigs and their leaves for September 15, 1912, plotted with relative transpiration for September 14 and 16. Exp. XIX.

and a new curve was made by connecting the points of intersection. Figure 23 gives these resulting curves one space below their actual position in order to avoid numerous intersections with the original curves. This method of obtaining curves comparable in time is based on the assumption that both the relative transpiration and the water content change at a constant rate between the successive readings.

DISCUSSION.

The curves resulting from the plotting of water content determinations of the two trees, I and II, are fairly consistent in general shape. In the case of the end twigs there is a minimum water content about noon followed by a subsequent rise. The decrease begins in one case about 8^h 30^m a. m. and in the other at 10^h 30^m a. m.; in the former case recovery is slower than in the latter. In the case of branches a meter from the ends it may be said that in general the slopes of the curves are in opposite directions to the slopes of the twig curves. The branches 2 meters from the ends behave differently in the two trees; in I_3 the curves are, in general, parallel to the I_2 curve, while curve III_3 is nearly flat or only slightly lower in the middle of the day than at the two ends of the day.

The curve of relative transpiration from experiment III (see p. 20) is plotted at the top of fig. 22. Although experiment III was performed in March and experiment XVIII in August, the conditions of soil moisture and evaporating power of the air at the two dates make it seem fair to compare the two sets of readings. The soil moisture at a depth of 30 cm., as given by weekly determinations made by Dr. Forrest Shreve, was 15 per cent on March 12 and 15.7 per cent on August 6. The summer rains which had intervened allowed for recovery from the extreme conditions of drought obtaining during the arid fore-summer. It may be, of course, that transpiration conditions were different in the two cases, but the probability is that they were much the same.

The following theory for the behavior shown is suggested. The relative transpiration rate increases until about 9^h 30^m a. m., when the demand for water made upon the plant by the evaporating power of the air exceeds the supply of water furnished by the transfer from the lower portions of the plant, and there results a slight drying out of the walls of the cells which form the true transpiring surface, both internal and external. This operates in two directions: the one, a lessening of the relative transpiration rate because the evaporating surface of the intercellular spaces has receded into the pores of the cell walls; the second, an assumption of the tensile state of the water in the vessels and cell walls, which results in a "pull" of water from the lower portions of the same stem and, as an immediate result of this, a transfer of water from lower to higher portions faster than the supply still farther down can stand, and therefore an increase in the size of the air bubbles in the vessels caused by the vacuum formed and hence a lower water content of the branches 1 meter down. Then 2 meters down the stem the effects might either lag behind those above or, being influenced by the same extra pull, might draw enough water from the portions still lower to keep up with the water content of the higher portions. Indeed, it would be impossible to predict the condition of any portion below the end or those very near the end. To return to the relative transpiration curve at $10^h\ 30^m$ a. m., it will be seen that the rate is now lessened as a result of the drying out of the tissues. This decreased transpiration gives the water supply a chance to catch up with the demand and a straight line of relative transpiration might be expected for a short time at least; but here a marked increase in the water content of the twigs is observed, accompanied by a decrease 1 meter below. This increase is accompanied also by an increase in the relative transpiration above, a natural result of the water films coming nearer the surfaces of the pores.

Experiment XIX differs from experiment XVIII in that the tree was in leaf and that the transpiration rates are more surely comparable with the water content readings. An inspection of the curves shows the following facts. The relative transpiration increases until about $9^h\ 30^m$ a. m., being accompanied by a decrease in the water content of the leaves and an increase in that of the twigs. Until $11^h\ 30^m$ a. m. the water content curves of leaves and twigs slant in opposite directions, after which time they run parallel.

Bearing in mind the theory mentioned in the preceding paragraph, the following points are evident: The increase in relative transpiration up to $9^h\ 30^m$ a. m. brings demand greater than supply, causing a decrease in the water content of the leaves; this in turn throws the water in the tissues into a tensile state, causing a rise of water content in the twigs by the water brought up from below. From $9^h\ 30^m$ a. m. to $10^h\ 30^m$ a. m. the transpiration rate is decreased, caused by the lowered amount of water present, but at the same time the water content of the leaves is slightly increased by a "pull" which lowers the water content of the twigs. By $10^h\ 30^m$ a. m. the transpiration and relative transpiration have been able to increase because of the rising water content of the leaves, but this water is gained at the expense of the twigs and it is not until after 12 noon that the lower portions can supply water fast enough for the water content of twigs and leaves to increase together. During the afternoon the relative rate fluctuates slightly from a straight line. From 4 p.m. to 5 p.m. the two transpiration curves go in opposite directions, so no conclusions from their comparison with the water content can be made.

In cases of this kind, where the tree is in leaf, the drying out is probably accompanied by a closure and subsequent opening of the stomata, as was mentioned in the section on stomatal behavior. The fall and rise in relative transpiration may occur more frequently than the readings at hourly intervals show. In fact, it is possible that here is an explanation for the well-known small fluctuations in relative transpiration observed when readings are taken at very short intervals. On page 6 is a brief description of the dying of twigs, branches, and finally limbs, which takes place during times of drought. This would happen when the tensile "pull," due to a deficit of water in the upper tissues, became great enough to cause a breakage of all or nearly all of the water columns. The twig would then quickly die for want of water, even though the lower portions remained alive.

On page 36 mention was made of the fact that there seems to be a relation between the amount of the maximum of actual transpiration immediately preceding the drop and the occurrence of the drop. This fact seems to accord with the theory given above, the maximum amount of actual transpiration for any tissue being a fairly constant factor dependent for each tissue upon its structure and water content, on the one hand, and upon the rate at which it receives water from lower tissues on the other.

DAILY COURSE OF LEAF TEMPERATURE.

METHOD.*

The method to be described is one commonly used by physicists in the determination of specific heat, heat of evaporation, etc., the underlying principle being that when two substances at different temperatures are placed in contact and external loss or gain of heat is eliminated, the amount of heat gained by the cooler body is equal to the amount lost by the warmer one. Or, expressed more concretely:

$$(A)(Sa)(Ta-Tc)=B(Sb)(Tc-Tb)+C(Tc-Tb) \qquad (1)$$

where

A = weight of substance at the higher temperature.
B = weight of substance at the lower temperature.
Sa = specific heat of substance A.
Sb = specific heat of substance B.
Ta = initial temperature (centigrade) of A.
Tb = initial temperature (centigrade) of B.
Tc = final temperature (centigrade) of the mixture.
C = number of calories required to raise calorimeter 1° C.

In the problem under consideration Ta, the unknown required, can be calculated since all the other factors can be found.

The finding of a substance with which to bring the leaf in contact offered some difficulties. An ideal substance would have to answer the following conditions: (1) a liquid in which the leaf may be completely immersed, which will come into intimate contact with the surfaces of the leaf, and will penetrate to the interior; (2) a material of low specific heat and low density, so that small differences of temperature in leaves may register large changes in a small amount of liquid; (3) a non-volatile material, in order to avoid

*A preliminary announcement of this method was made in the Johns Hopkins Univ. Cir., Feb. 1912.

errors from loss of weight by evaporation while the experiment is in progress; (4) a substance which when mixed with water shows no temperature changes (the high water content of most leaves makes this important); (5) a substance which is miscible with water; (6) a substance with which substances in the leaf produce no chemical changes accompanied by heat; (7) a substance obtainable in the market at a reasonable price; (8) a good conductor of heat, in order that the final temperature may be reached quickly.

This ideal substance has not yet been found, but there are several materials which satisfy most of the important conditions. Water, the substance most commonly used by physicists in work of this kind, is ruled out by the first condition, as it collects in drops on the surfaces of hairy leaves, leaving bubbles of air on the leaf which will act as non-conductors of heat. The second condition also helps to rule out water. The third condition might be overcome by using a large enough quantity to reduce to a minimum the percentage of error from evaporation were it not for the fact that the specific heat is so high that a large amount of water requires a larger amount of leaves than it is practicable to use in the case of small-leaved plants. Alcohol answers all the conditions with the exception of the third. It is so volatile that troublesome precautions would have to be taken to avoid a very appreciable error. Common glycerine causes a rise of temperature when mixed with water and therefore can not be used.

Turpentine, the substance used in the following experiments, was found: (1) to penetrate the leaves immediately, so that they appeared "cleared" under the microscope; (2) to have a satisfactory combination of specific heat and density for the use of small amounts; (3) not to be more volatile than the accuracy of the other factors in the experiment as a whole required; (4) to show no temperature change when mixed with either large or small amounts of water; (6) to cause the release of no measurable amount of heat when the leaves of three species of plants were immersed in it; (7) to be easily obtainable in the market at a reasonable price; but samples from different distillations show variable specific heats; (8) to have a high enough heat conductivity for equilibrium temperature to be reached in 2 or 3 minutes. Turpentine is thus seen to answer fully all of the conditions except the fifth and seventh. In regard to the fifth condition it was found that although turpentine does not mix with water, yet the absence of small bubbles under the microscope, when the leaves used had been immersed in turpentine, led to the assumption that no air-spaces of sufficient size to influence the results existed. As for the seventh condition, the disadvantage arising from the irregularity in the specific heat of turpentine can be overcome by a little extra labor, for the specific heat can be determined accurately to the second decimal place with the apparatus used in the rest of the experiment or with an ordinary double calorimeter. Samples taken at various times from the same purchase show an agreement in specific heat, and so if sufficient quantity be obtained at the beginning of the experiment to last for a set of readings the specific heat need be determined but once.

Loss or gain of heat from the surrounding air by the vessels may be prevented by two methods: (1) By surrounding the vessel with a non-conducting material; (2) by cooling the liquid to a temperature the same number of degrees below room temperature that the final temperature will reach above room temperature, thus eliminating the error on the theory that the heat lost to the air is equal to the heat gained from the air in the same length of

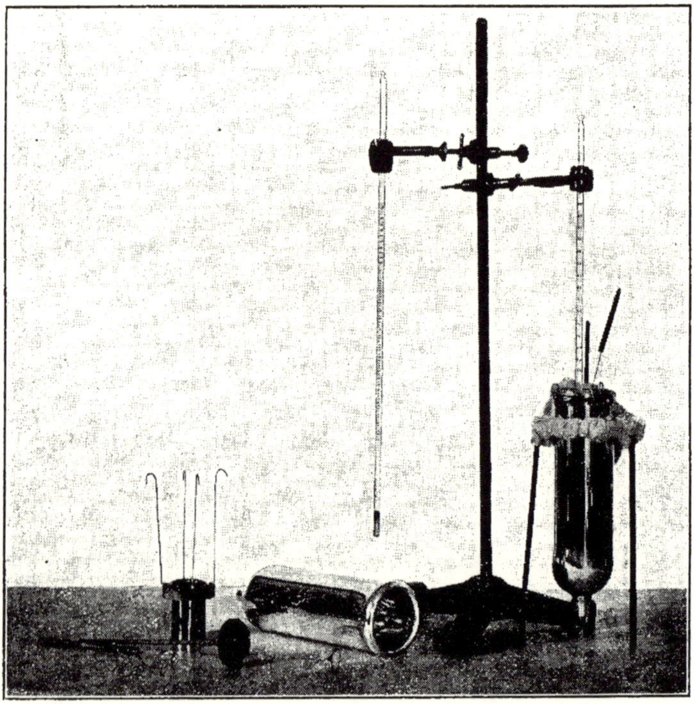

FIG. 24—Apparatus for determination of leaf-temperature.

time. The latter requires a previous knowledge of the temperatures and amounts, and this is wasteful of leaf material. The former method is difficult to carry out in its ideal form, but by keeping the temperatures near air temperature, and the use of a predetermined "curve of cooling," very accurate results may be obtained with well insulated vessels.

The apparatus used was made for the purpose and is merely an adaptation of the ordinary calorimeter with a jacket. From fig. 24 it will be seen that the apparatus consists of a calorimeter suspended nearly to the bottom of a silvered Dewar vacuum beaker. The calorimeter cup is made of very thin copper, nickel-plated, and polished to a bright surface on the outside. A wooden cover, with holes for thermometer and stirrer, fits snugly in the top, while around the outside is a wooden rim which fits closely to the sides of the beaker, giving no chance for entrance and exit of air. This wooden rim is suspended from the top of the Dewar beaker by means of metal wires. Nowhere does the calorimeter cup touch the beaker or the metal wires. Cotton is used for a cover at the top, as an extra precaution. A certified one-tenth degree thermometer was used in the experiments.

That the apparatus is well protected from exchange of heat with the surrounding air, when temperatures within 5 degrees of the outside air are used, may be seen from the fact that when the calorimeter cup was filled with turpentine at 5 degrees below air temperature, then put in its position in the beaker and left in the sunshine in August for 15 minutes, a rise of only 0.2 degree occurred. Further, the regular rate of warming or cooling, as indicated by all the experiments, showed that a curve of cooling could be depended upon for use in the small ranges of temperature which obtained in the experiments.

In order to find the temperature of a leaf by immersing it in turpentine, the following quantities must either be known already or must be found: (1) the specific heat of the turpentine used; (2) the amount of heat absorbed by the calorimeter; (3) the specific heat of the leaves.

(1) The specific heat of the turpentine was determined according to the customary method followed by physicists of using a double calorimeter of capacity of over 500 c.c. Three determinations were made which differed in the third decimal place only, results being 0.409, 0.410, 0.411.

(2) The amount of heat absorbed by the calorimeter system was determined as follows. The calorimeter cup with thermometer and stirrer in place was filled with a weighed amount of turpentine to a marked level, this marked level being used subsequently as a guide for the level of the liquid after the leaves were put in. The vessel with the turpentine was cooled to about 12° C., then weighed and put in place in the Dewar beaker, where it was allowed to come to partial equilibrium. When the rate of warming had slowed down to about one-tenth degree a minute a weighed amount of turpentine at air temperature was quickly added, the mixture filling the vessel about three-fourths full. The exact temperature of each liquid was taken immediately before the mixing. The cover was instantly replaced and the mixture stirred gently and continuously. The time of the transfer occupied 2 to 5 seconds. The time and temperatures were then read at intervals of half-minutes until equilibrium was reached or until the rate of change equaled the rate of change which had been previously shown to exist for the temperature obtaining.

(3) In order to determine the specific heat of the identical leaves used each time and to get them in the same condition in which they were used for the

temperature determination, *i. e.*, in the turpentine, their specific heat was determined last; meanwhile readings for their initial temperature were taken as follows.

Turpentine at air-temperature was placed in two calorimeters and the whole allowed to come to equilibrium on a shaded stand near the tree to be used. Leaves were quickly severed from their twigs by means of wooden-tipped forceps and then placed on white glazed paper, from which they were transferred with all possible speed to one of the calorimeter cups, the other being kept for a control. The contents of both calorimeters was kept in constant slow motion by use of the stirrers. Temperatures were read every half minute before and after the introduction of the leaves.

A determination of the specific heat of the leaves was made by heating turpentine to about 30° C., then placing it, in the same careful way as above, into the turpentine and leaves just used in the temperature readings. Temperatures of the contents of both liquids were taken accurately just before and after the mixing. The specific heat of the leaves was calculated from the equation,

$$[A(Sa)+L(Se)+C][Ta-Tc]=B(Sb)(Tc-Tb) \qquad (2)$$

where

A = weight of turpentine containing leaves.
B = weight of turpentine added to A, and leaves.
Sa, Sb = specific heat of turpentine.
L = weight of leaves.
Se = specific heat of leaves.
Ta = initial temperature of A and L.
Tb = initial temperature of B.
Tc = temperature of mixture.
C = heat equivalent of the calorimeter.

After the value of Se had been found it was substituted in the equation

$$[A(Sa)+C]\ [Ta-Tc]=L(Se)\ (Tc-Tb) \qquad (3)$$

where

A = weight of turpentine prepared to receive leaves from the plant.
L = weight of leaves.
Ta = initial temperature of turpentine.
Tb = initial temperature of leaves.
Tc = temperature of mixture.
Se and C = same as in previous cases.

Since Tb is now the only unknown quantity in equation 3, the original temperature of the leaves can be easily calculated.

Following is an example of the calculation for one determination.

TABLE 22.—*Temperature readings for the determination of leaf-temperature.*

Calorimeter C.		Calorimeter T.		Air temperature.
Time.	Temperature	Time.	Temperature.	
READINGS FOR LEAF-TEMPERATURE, EQUATION 3.				
	° C.		° C.	
$10^h\ 50^m$ a.m.	18.58	$10^h\ 50.5^m$ a.m.	18.32	....
11 10 a.m.	18.58	11 10.5 a m.	18.32	....
Opened for 1 min.				
$11^h\ 12^m$ a.m.	18.58	11 12.5 a.m.	18.32	....
11 30 a.m.	18.58	11 30.5 a.m.	18.32	....
12 22 p.m.	18.70	12 22.5 p.m.	18.46	....
12 23 p.m.	18.70	12 23.5 p.m.	18.46	....
12 24 p.m.	18.70	12 24.5 p.m.	18.46	....
12 25 p.m.	18.70	12 25.5 p.m.	18.46	19.0
Leaves put in.				
$12^h\ 26^m$ p.m.	18.90	12 27 p.m.	18.46	....
12 26.5 p.m.	19.10	12 28.5 p.m.	18.46	....
12 28 p.m.	19.16	12 29.5 p.m.	18.46	....
12 29 p.m.	19.19	12 30.5 p.m.	18.46	....
12 30 p.m.	19.20	12 31.5 p.m.	18.46	....
12 31 p.m.	19.20	12 32.5 p.m.	18.46	....
12 32 p.m.	19.20	12 33.5 p.m.	18.46	....
12 33 p.m.	19.20	12 34.5 p.m.	18.46	....
12 34 p.m.	19.20	12 35.5 p.m.	18.46	....
12 35 p.m.	19.20	12 36.5 p.m.	18.46	....
12 37 p.m.	19.20	12 37.5 p.m.	18.46	....
12 40 p.m.	19.20	12 40.5 p.m.	18.46	....
12 50 p.m.	19.18	12 50.5 p.m.	18.47	19.0
READINGS FOR SPECIFIC HEAT OF LEAVES, EQUATION 2.				
	° C.		° C.	
$1^h\ 18^m$ p.m.	15.13	$1^h\ 18.5^m$ p.m.	15.10	....
1 19 p.m.	15.13	1 19.5 p.m.	15.10	....
1 20 p.m.	15.13	1 20.5 p.m.	15.10	....
1 22 p.m.	15.13	1 22.5 p.m.	15.10	17.7
Turpentine *B* put in.				
$1^h\ 24^m$ p.m.	21.60	1 24.5 p.m.	15.10	....
1 25 p.m.	21.60	1 25.5 p.m.	15.10	....
1 26 p.m.	21.60	1 26.5 p.m.	15.10	17.7
1 40 p.m.	21.55	1 40.5 p.m.	15.15	....
1 57 p.m.	21.35	1 57.5 p.m.	15.20	18.8

Readings for leaf-temperature determination (equation 3).

Weight of calorimeter *C*, stirrer, thermometer, etc.	121.08 gm.
Weight of C, etc., plus turpentine *A*	141.74
∴ Weight *A*	20.66
Weight *C* plus *A* plus leaves	143.97
∴ Weight of leaves	2.23
Temperature *Ta*	18.70°
Temperature *Tc*	19.20°
Temperature of air	19.0°

Readings for specific heat of leaves (equation 2).

Weight C, plus *A*, plus leaves, plus turpentine *B*	158.48 gm.
∴ Weight *B*	14.51
Temperature *Ta*	15.30°
Temperature *Tc*	21.60°
Temperature *Tb*	28.84°
Temperature of air	17.7°

Proper substitution in equation 2 gives:

$$[(20.66)(.41)+(2.23)(X)+6.80][6.3]=(14.51)(7.24)$$

$$\therefore \text{ specific heat of leaves, } X=.64.$$

Proper substitution in equation 3 gives:

$$[(20.66)(.41)+6.80][.50]=(2.23)(.64)(X) \therefore X=5.3.$$

Therefore the temperature of the leaves was

$$19.2°+5.3°=24.5°, \text{ or } 5.5° \text{ above air-temperature.}$$

EXPERIMENTATION.

Experiment XX.

A series of leaf-temperature determinations was made from trees *D* and *E*. The samples were taken every 2 hours, a period of 2 to 3 minutes elapsing between the cuttings. The general appearance of the trees indicated that tree *E* was nearer the stage of dropping its leaves than was *D*.

Table 23.—*Temperatures for leaves from two trees with the corresponding air-temperatures Experiment* XX.

Date.	Time.	Temperature.			Difference from air.	
		D	*E*	Air.	*D*	*E*
1912.		° *C.*	° *C.*	° *C.*		
Sept. 16...	$9^h\ 00^m$ p.m.	26.7	27.6	27	−1.3	+0.6
	11 59 p.m.	22.9	24.8	25	−2.1	−0.2
				D. *E.*		
Sept. 17..	5 00 a.m.	19.6	16.0	20 19	−0.4	−3.0
	8 00 a.m.	22.8	24.3	25.5 24.8	−2.7	−0.5
	10 00 a.m.	32.9	35.8	33.5	−0.6	+2.3
	12 00 m.	45.9	44.6	37.5	+8.4	+7.1
	2 00 p.m.	38.0	43.7	39.0	−1.0	+4.7
				D. *E.*		
	4 00 p.m.	35.1	35.9	39.5 39.0	−4.4	−3.1
	6 00 p.m.	29.1	27.9	32.5	−3.4	−4.6

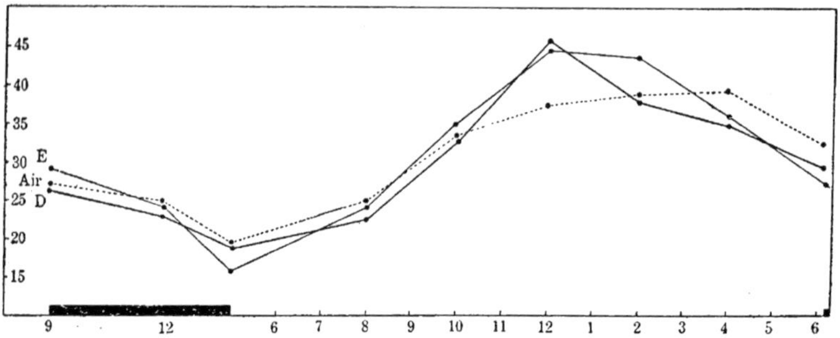

Fig. 25.—Graph for leaf-temperature from two adult trees, plotted with air-temperatures. Exp. XX.

DISCUSSION.

From an inspection of the leaf temperature curves in figure 25 it is seen that in spite of small hourly differences the two trees show much the same behavior as compared with the air temperature. In general the temperature curves for the leaves are below the air temperature curves at night and in early morning until about 10 o'clock, when the lines cross the air temperature lines and then remain above until some time between 12 and 1 o'clock, when they again cross and go below air temperature. When compared with the water content curves, no actual hourly relationship appears, there being merely the general relationship that the lowered water content during the day is accompanied by a rise in the leaf temperature. Since both humidity and temperature enter into evaporation control and since the amount of heat received from the sun and air is accumulative, too many factors must be considered to make conclusive deductions. The unknown factor is the balance between the heat received and that used in evaporation. The increase of transpiration for the morning does not keep pace with the increasing amount of heat energy received, and hence it is to be expected that less of the heat received by the leaves would be used by the water in evaporation and therefore more would be available for raising the temperature of the tissues. The rise in leaf temperature shows, at least, that transpiration and relative transpiration are not lessened during the middle of the day because the sun's energy is being used in other ways, *i. e.*, in photosynthesis, for the heating of the tissues shows that there is heat energy to spare.

It will be noted that the leaf temperature begins rising above the air temperature about the time the closure of stomata commences, and begins sinking with the opening of the stomata. While the above temperature curves show no conclusive evidence of a connection with the dip and rise in relative transpiration in the morning, they present no negative evidence for the application of the theory of drying, since the ratio of the energy received to the amount of evaporation can not be determined. It is interesting to note that the lowered evaporation rate during the middle of the day is not great enough to keep the leaf temperature down to air temperature, and that, while there is some evidence of the desiccation theory, the unknown amount of accumulated heat energy makes this evidence inconclusive.

DAILY COURSE OF TRANSPIRATION UNDER CONDITIONS OF HIGH AND LOW EVAPORATION.

If the occurrence of the dip is caused by a failure of the ratio of water supply to demand, to equal unity, then on *a priori* grounds a plant showing the dip under conditions of high evaporation either ought not to show it at all under conditions of low evaporation or to a much less extent. This was tested to some extent in experiment VIII, where a plant which had previously shown a dip was run in a room of the laboratory. Although the plant showed no dip under these circumstances, the absence of sunlight makes the experiment inconclusive in this respect. Indeed, if the cause of the dip is connected with the rate of production of metabolic water, the presence or absence of sunlight would be a disturbing factor. Consequently another experiment was planned in which the same plant would run under natural conditions of low humidity and sunlight at one time, and at another time in the sunlight, but under conditions of high humidity.

METHOD.

Although potted plants have been seen to show the dip much less pronouncedly than trees, nevertheless the impossibility of getting the entire tree under controlled conditions made the use of potted plants seem necessary; so in September 1912 another attempt was made to obtain transplanted seedlings. A dozen or so plants, varying in age from two to fifteen years, were very carefully removed from the soil and transferred to pots. These were placed in the green-house, where they remained until March 1, when the two surviving ones were transferred to the roof of the laboratory and there left until the time of the following experiment.

Aluminum shells of the type devised by Ganong* were used to seal the pots. The tops of the pots were first sealed with plastocene, over which was fastened a rubber sheet, according to Ganong's instructions. Since the plants lose only a small amount of water per hour, the aluminum shells with their rubber tops made the danger of error from dust particles much less. Before each weighing the shells and rubber were dusted off with a camels-hair brush. The balances used were sensitive to milligrams, but readings were taken only to the nearest 10 mg.

EXPERIMENTATION.

Experiment XXI.

The subjects of this experiment were two potted plants, Nos. 6 and 7, whose previous history is given in a former paragraph. The plants were sealed on the night of April 1. Plant No. 7 was given water immediately before being sealed, and No. 6 had received no water for two days.

On April 2 the two plants with two atmometers were placed in an exposed position near the laboratory and the hourly rate of transpiration was measured. Each plant with its atmometer was carried into the laboratory for the weighings, care being taken each time to return plant and atmometer to their original positions in the open. During the entire morning the sky was partly cloudy and the sun was intermittently obscured. The wind velocity was extremely high all day. The results appear in table 24, where the numbers given are the actual losses from the plants, since the area was not obtained. An examination of these results shows that No. 6 had an early drop in both actual and relative transpiration, which was followed by

*Bot. Gaz., 41, 212, 1906.

a lessening in the rate of decrease but not by an actual increase. The occurrence of the maxima of actual and of relative transpiration was 4 and 6 hours, respectively, before the time of maximum evaporation for the day. No. 7 behaved differently in that its maximum of actual transpiration coincided with the maximum for evaporation and no tendency toward a break in the acceleration appeared. The maximum for relative transpiration occurred four hours earlier than the transpiration maximum, and while a

TABLE 24.—*Transpiration under three different evaporation conditions. Experiment XXI.*

Date.	Time.	E	Plant No 6.		Plant No. 7.		Temperature.	Humidity.	Remarks.
			T	$\frac{T}{E}$	T	$\frac{T}{E}$			
							° F.	p. ct.	
April 1...	7^h 08^m p.m.	...	....	.....	....	.....	....	..	Cloudy. Wind high, leaflets open.
April 2...	6 30 a.m.	1.99	0.07	0.035	0.05	0.025	64.5	60	
	9 03 a.m.	2.28	.24	.105	.24	.105	66	58	Sunlight intermittent. Leaflets, 35°. N› change in conditions.
	10 05 a.m.	3.37	.47	.139	.49	.145	67	57	
	11 05 a.m.	4.52	.38	.086	.55	.122	68	55	
	12 05 p.m.	4.58	.37	.081	.58	.127	72	50	Do.
	2 20 p.m.	5.44	.31	.057	.60	.110	72.5	50	
	3 55 p.m.	4.97	.28	.056	.29	.058	....	48	Do.
	7 15 p.m.	4.28	.20	.047	.24	.056	....	62	Dark at close.
April 4 ..	6 45 a.m.	....	....	.....	....	.....	....	..	
	8 10 a.m.	2.11	.18	.085	.18	.085	53	50	
	9 05 a.m.	2.21	.39	.176	.32	.145	61	48	Leaflets about 20° apart after 7 a.m.
	10 05 a.m.	2.48	.40	.161	.41	.165	70	46	Sun bright and air almost still all day.
	11 05 a.m.	3.19	.35	.110	.55	.172	73.5	42	
	12 05 p.m.	3.80	.44	.116	.61	.161	79	40	
	1 30 p.m.	4.53	.44	.097	.66	.146	88	35	
	2 35 p.m.	4.58	.44	.096	.63	.138	88	35	
	4 15 p.m.	4.52	.39	.086	.55	.122	....	36	
	7 15 p.m.	3.02	.38	.126	.29	.096	77	38	
April 11..	6 36 a.m.	....	....	.....	....	.....	....	..	Sun bright all day. Moist chamber protects plant and atmometer from full force of a strong wind.
	8 03 a.m.	1.47	.03	.020	....	.....	....	..	
	9 06 a.m.	1.15	.01	.009	....	.....	55	70	
	10 03 a.m.	1.26	.08	.063	....	.....	55	80	
	11 04 a.m.	1.49	.12	.081	....	.....	56	80	Leaflets 160° apart until 4 p.m., after which 90° apart.
	11 59 a.m.	1.31	.12	.092	....	.....	58	85	
	12 58 p.m.	1.40	.14	.100	....	.....	58	80	
	2 08 p.m.	1.40	.17	.121	....	.....	65	75	
	4 07 p.m.	1.05	.10	.095	....	.....	65	50	
	7 30 p.m.	3.98	.19	.048	....	.....	70	50	
April 11.	7 45 a.m.	....	....	.....	....	.....	....	..	
	8 55 a.m.	4.96	....	.....	.50	.101	71	35	Sun bright and wind velocity very high all day.
	9 49 a.m.	4.10	....	.....	.63	.154	72.5	32	
	10 57 a.m.	5.43	....	.....	.75	.138	74	30	Leaflets 10° to 15° apart during period of experiment.
	11 53 a.m.	5.68	....	.....	.82	.144	80	27	
	12 52 p.m.	5.91	....	.....	.82	.139	83	24	
	2 00 p.m.	6.04	....	.....	.90	.149	83	22	
	4 00 p.m.	5.86	....	.....	.87	.148	81	22	
	7 27 p.m.	3.58	....	.....	.51	.142	....	23	

drop and rise occur in these rates it will be seen from a comparison with the transpiration and evaporation rates that they are caused by a decrease in the evaporation rate which is not followed by the plant, and hence that this dip and rise are in no sense the kind under investigation in this paper.

The intermittent cloudiness of April 2 made it advisable to repeat the experiment on a clear day, and consequently this was done on April 4.

The seals of the pots were not broken in the meantime. This day proved to be continuously clear, with practically no air movement. The results appear in table 24. No. 6 showed a drop and rise in both actual and relative transpiration, being of much less magnitude in the former case than in the latter. No. 7 behaved very much as on April 2, the maximum for actual transpiration coinciding with that of evaporation and the maximum for relative transpiration occurring earlier.

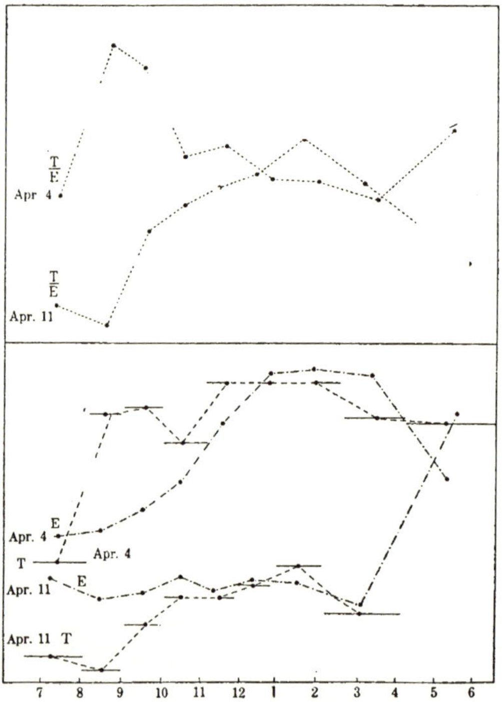

FIG. 26.—Graphs for transpiration of plant No. 6, under two different conditions of evaporation. Exp. XXI.

Under conditions of high evaporation thus far, No. 6 has shown a morning drop and rise and No. 7 has not. The next plan was to put No. 7 under conditions of still higher evaporation and No. 6 under conditions of lowered evaporation. On April 11 the natural conditions provided greater evaporation than was obtained on either of the previous days, as the sun was bright and the wind velocity extremely high during the entire day. In order to secure the desired conditions for No. 6, a wooden box large enough to hold plant, atmometer, a Lambrecht's polymeter, and a thermometer was lined with wet towels; an opening was left in the side and top of the box for the entrance of sunlight and air. The towels were wet frequently and the box was turned as the day advanced, so that no shadow ever fell on plant or atmometer. The rapid evaporation caused an unavoidable lowering of temperature within the box. The polymeter had to be so placed that the humidity readings are probably a little too low.

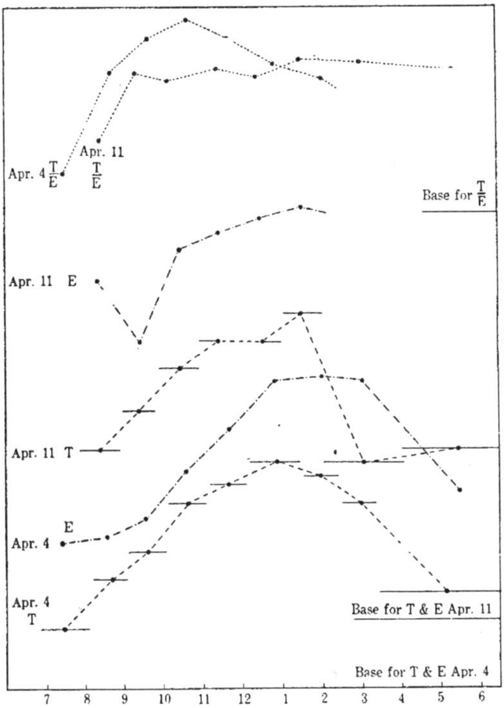

FIG. 27.—Graphs for transpiration of plant No. 7, under two conditions of evaporation rate. Exp. XXII.

These results also appear in table 24. They show that the evaporation conditions for No. 6 were greatly reduced and that the actual transpiration was greatly cut down. The maxima for both relative and actual transpiration are coincident with the evaporation maximum. Thus, without doubt, the drop and rise do not occur under the conditions of lessened evaporation, but do occur in this same plant under more strenuous evaporation conditions. In the case of No. 7, the evaporation conditions were more severe than on April 4, but the difference is by no means so marked as was the case with No. 6. The actual transpiration amounts are higher than before, showing that the plant was capable of giving off more water than on April 4, but the relative transpiration is on the whole lower. Furthermore, a flattening of the curve of actual transpiration occurs between 11 and 1 o'clock, and the relative curve shows a corresponding dip and rise. Evidently the limit of maximum water loss for this plant, under the existing conditions of soil moisture and water content of tissues, is being reached, and it is probable that the more severe conditions of midsummer would cause a more pronounced drop. It will be noted that the maximum transpiration amount of No. 7 on April 11 seems to bear somewhat the same relation to its limit as does that of No. 6 on the same date, although the evaporation conditions were very different. The facts that No. 7 received water just preceding the sealing and that No. 6 had been without water for 48 hours make these results agree in a general way with the results of experiment X.

DISCUSSION OF LITERATURE.

Lloyd* has taken simultaneous readings of transpiration, evaporation, water content of leaves, and stomatal movement, but, while his readings doubtless answer satisfactorily the question which he had in mind, they are of no avail for discussion in this paper because of a haziness in the sky which he records as existing between 8 a. m. and 10 a. m.

Livingston and Brown,† find in general that green plants, in the desert of Arizona, "exhibit a marked fall in foliar moisture content by day and a corresponding rise by night," and they give as the probable cause of this, "incipient drying brought about whenever the ratio of water loss to water supply in the leaves is rendered less than unity." Unfortunately the above paper had not appeared when the greater part of the present work was carried out, and so it could not be taken into account in the planning of the experiments. Livingston and Brown have evidence which leads them to suggest that small-leaved xerophytes show a "higher water content by day than by night." The present work shows that this is not true in the case of *Parkinsonia,* for in addition to the hourly curve of leaf moisture given, three different tests, made on trees on different days, show the lowest daily water-content at midday. On the whole it may be said that the present paper offers additional evidence for Dr. Livingston's theory of "incipient drying" of leaves and shows that the same phenomenon takes place also in stems.

---

* Lloyd, F. E., The relation of transpiration and stomatal movements to the water content of the leaves of *Fouquieria splendens.* Plant World, vol. xv, p. 1, 1912.

† Livingston, B. E., and Brown, W. H., Relation of the daily march of transpiration to variations in the water content of foliage leaves. Bot. Gaz., vol. LIII, p. 311, 1912.

GENERAL SUMMARY.

(1) The relative transpiration rates of the plants used differ according to the previous environmental history of the plant; so that conclusions regarding the actual transpiration rates of plants *in situ* can not be drawn from the measurement of water losses from potted plants. Different branches of a single tree differ in relative transpiration rate no more than do the relative transpiration rates of the same branch on different days, and, furthermore, relative rates of different branches agree as well as do the rates of different potted plants taken on the same day or the rates of the same potted plant on different days. Therefore the transpiration behavior of an adult tree may be known better from the measurement of small branches than from the measurement of potted plants.

(2) The maxima for relative transpiration of potted plants of *Parkinsonia microphylla* were found to vary directly with the soil moisture, and slight evidence was obtained for the same variation in the case of plants *in situ*.

(3) The maxima of relative transpiration varied with the structure of the tissues in the order to be expected from their anatomical structures.

(4) The actual transpiration of adult trees and of young seedlings growing in the open shows a maximum, which occurs 2 to 3 hours earlier than the maximum of evaporation for the day. This maximum is followed by an abrupt drop and a subsequent rise, the rise being much more pronounced in the case of branches in leaf. This rise is in general great enough to appear as a rise in the relative transpiration. In the case of green-house-grown potted plants of the same species the early maximum appears in the relative transpiration, but not always in the actual transpiration, and a drop appears in various forms all the way from a distinct drop in the actual transpiration to only a slight flattening in the slope of the relative transpiration curve. The early maximum does not appear even in the relative rate when readings are taken in the shade.

(5) The curve of stomatal behavior follows the relative transpiration curve in such a manner that the existence of an interrelation is evident. When potted plants having a soil moisture of 14 per cent and 3 per cent are compared in respect to stomatal behavior and transpiration, a relation between the amounts of transpiration and the size of stomatal openings is seen to exist, but the amounts are not even approximately proportional to the linear dimensions of the stomatal openings. No conclusions regarding the action of cause and effect can be drawn from the measurements taken.

(6) Water content determinations of leafless twigs and of branches a meter distant from the twigs show an inverse relation, while the curve for the twigs follows the relative transpiration curve. Curves from further determinations of water content of twigs and their leaves show a relation to transpiration and relative transpiration curves, from which is offered the theory that the drop in relative transpiration and in actual transpiration is due to a slight drying out of the tissues. This theory is further strengthened by

experiment XXI, in which it appears that a plant which shows the dip and rise under conditions of high evaporation does not show them under conditions of lower evaporation. The water content of twigs and leaves suggests also a theory for the cause of the rise succeeding the drop, which is based on Dixon's theory for ascent of sap in trees.

(7) *Parkinsonia* trees in sunlight show hourly changes in the relative transpiration rate, the amount of opening of stomata, in water content of leaves and twigs, and in leaf temperature, and these have evident interrelations which are held to be governed by the ratio of the demand to the available supply of water.

Since, as is well recognized, the transpiration-absorption water-balance is probably the most vital factor governing the occurrence and distribution of plants in desert and semi-desert regions, the facts brought out in this paper are obviously connected with the success of *Parkinsonia* as a desert perennial. The story of the responses of the adult tree to the coming on of drought conditions is as follows. First, the leaflets begin closing earlier each day until finally they remain open only a few minutes at dawn and at twilight. Second, the transpiration amount is lessened with the drying out of the soil. Third, the leaflets drop off and later the rachises. Fourth, the twigs and small branches begin to die until finally, when extreme drought conditions prevail, whole limbs are lost without injury to the vitality of the plant, and thus the tree passes through the drought period in spite of the large amount of evaporating surface still exposed at the time of the falling of the leaves. In addition to the seasonal responses, the tree has a daily response which consists in a closure of leaflets, followed several hours later by a lessening of actual transpiration rate, while the evaporating power of the air is still increasing. This decrease is accompanied by a closure of stomata, a lowered water content of leaves and twigs, and a slight rise in leaf temperature. The drop is followed by a rise, but in general the maximum transpiration is not again reached for that day. That seedlings do not respond so readily to seasonal changes is to be expected from the size and structure of their leaves, since these differ little from those of green-house-raised plants. Seedlings are frequently found in drought seasons with dead leaves attached, showing that the leaves were killed outright without falling. Perhaps the more mesophytic type of leaf can not endure the "incipient drying" to so great an extent as the more xerophytic type found on the adult trees. It is hoped that the studies now in progress will throw more light on the variations of transpiration and root absorption with soil moisture.

Spotted something that isn't quite right?

Let us know and we'll fix it for future reprints. And as a thank you for your diligence, we'll send you a free Sothis book of your choice.

publisher@sothis.press

This page intentionally left blank